Springer Tracts in Advanced Robotics 103

Editors

Prof. Bruno Siciliano
Dipartimento di Ingegneria Elettrica
e Tecnologie dell'Informazione
Università degli Studi di Napoli
Federico II
Via Claudio 21, 80125 Napoli
Italy
E-mail: siciliano@unina.it

Prof. Oussama Khatib
Artificial Intelligence Laboratory
Department of Computer Science
Stanford University
Stanford, CA 94305-9010
USA
E-mail: khatib@cs.stanford.edu

For further volumes:
http://www.springer.com/series/5208

STAR (Springer Tracts in Advanced Robotics) has been promoted under the auspices of EURON (European Robotics Research Network)

Héctor M. Becerra · Carlos Sagüés

Visual Control of Wheeled Mobile Robots

Unifying Vision and Control in Generic Approaches

Springer

Dr. Héctor M. Becerra
Department of Computer Science
CIMAT (Centro de Investigación
en Matemáticas, A.C)
Jalisco S/N, Col. Valenciana, 36240
Guanajuato, Gto.
Mexico
E-mail: hector.becerra@cimat.mx

Prof. Carlos Sagüés
Instituto de Investigación en
Ingeniería de Aragón
Universidad de Zaragoza
Calle María de Luna 1
50018 Zaragoza
Spain
E-mail: csagues@unizar.es

ISSN 1610-7438 ISSN 1610-742X (electronic)
ISBN 978-3-319-35907-6 ISBN 978-3-319-05783-5 (eBook)
DOI 10.1007/978-3-319-05783-5
Springer Cham Heidelberg New York Dordrecht London

Printed on acid-free paper

Springer is part of Springer Science+Business Media (www.springer.com)

From Héctor M. Becerra:
To my lovely wife Noemi, my beloved son David and to my great family, specially to my parents Graciela and Javier.

From Carlos Sagüés:
To my darling wife Isabel, and sons Javier and Rubén. To my family, specially to my mother Basi, my sister Regina and my mother-in-law Maribel.

Foreword

Robotics is undergoing a major transformation in scope and dimension. From a largely dominant industrial focus, robotics is rapidly expanding into human environments and vigorously engaged in its new challenges. Interacting with, assisting, serving, and exploring with humans, the emerging robots will increasingly touch people and their lives.

Beyond its impact on physical robots, the body of knowledge robotics has produced is revealing a much wider range of applications reaching across diverse research areas and scientific disciplines, such as: biomechanics, haptics, neurosciences, virtual simulation, animation, surgery, and sensor networks among others. In return, the challenges of the new emerging areas are proving an abundant source of stimulation and insights for the field of robotics. It is indeed at the intersection of disciplines that the most striking advances happen.

The Springer Tracts in Advanced Robotics (STAR) is devoted to bringing to the research community the latest advances in the robotics field on the basis of their significance and quality. Through a wide and timely dissemination of critical research developments in robotics, our objective with this series is to promote more exchanges and collaborations among the researchers in the community and contribute to further advancements in this rapidly growing field.

The monograph by Héctor M. Becerra and Carlos Sagüés provides an extensive tract on visual control of wheeled mobile robots by resorting to a rigorous control theory formalism. The approach is based on the use of the epipolar geometry and trifocal tensor concepts. Stability in the closed loop, region of convergence and robustness against parametric uncertainty are of concern for the visual servoing technique presented in this work, which relies on a monocular vision system as the unique sensor. The resulting schemes are valid for vision systems obeying a central projection model, i.e., conventional cameras, catadioptric systems and some fisheye cameras, so that the well-known visibility constraint problem can be avoided with an adequate sensor.

Rich of practical examples and simulation/experimental results, this volume constitutes a very fine addition to STAR!

Naples, Italy
February 2014

Bruno Siciliano
STAR Editor

Preface

This book is focused in the formal treatment of some aspects of control theory applied to the problem of vision-based pose regulation of wheeled mobile robots. We deal with this problem in such a way that vision and control are unified to achieve stability of the closed loop, a large region of convergence (without local minima) and good robustness against parametric uncertainty. We present and evaluate experimentally different control schemes that rely on a monocular vision system as unique sensor. The schemes are valid for vision systems obeying a central projection model, i.e., conventional cameras, catadioptric systems and some fisheye cameras, so that the visibility constraint problem can be avoided with an adequate sensor. In the described schemes along the book, a minimum set of visual measurements are taken from a geometric constraint imposed between image features. Basically, we exploit the epipolar geometry and the trifocal tensor given that they are general and robust geometric constraints, similar to the homography model, which could be used for particular situations.

The book initiates with a general introduction specifying the context of the work. We have included a complete review of the state of the art on visual control, first from a general point of view where the control is carried out in 6 degrees of freedom, and then, in the context of wheeled mobile robots. Additionally, we introduce the mathematical modeling of the robotic system, the model of central cameras based on the sphere camera model and the models of the visual measurements used for feedback, which are taken from a multi-view geometric constraint.

The epipolar geometry has been exploited in the literature on visual control of mobile robots in a typical framework of two views, where a remaining longitudinal error is corrected typically by switching to a different visual information and control approach. We present a new control strategy performed in two steps to achieve position and orientation correction, which also corrects longitudinal error by exploiting the epipolar geometry that relates a third image with the target image. Thus, the presented epipolar-based control is a robust control law that takes advantage of the information provided by three images through their pairwise epipolar geometries. The use of a third image allows to unify the control scheme in only one type of image-based controller for the whole task. The well-known sliding mode control technique is used to achieve good performance in the presence of singularities, typically present in previous approaches, and good robustness against uncertainty in camera calibration parameters and image noise.

In our epipolar-based control, we appreciate the benefits of using three images instead of the typical two-view based approach. The trifocal tensor is a geometric constraint that intrinsically integrates the rich information of three views, however, it has been little exploited for visual control. Given that the trifocal tensor is an oversized measurement with respect to the robot's state, we have determined a suitable two-dimensional error function in order to ensure stability of a square control system. This image-based control scheme uses direct feedback of the elements of a simplified trifocal tensor: the one dimensional version. This trifocal tensor reduces the dimension of the image space from two to only one dimension, which is the bearing angle of the visual features. Although this scheme is also valid for conventional images, it particularly exploits the properties of omnidirectional images to preserve bearing information despite their high radial distortion. A sliding mode control law with direct feedback of the one-dimensional trifocal tensor results in a visual control scheme that does not require specific calibration.

Although the epipolar-based control and the trifocal-tensor-based control introduced in the previous paragraphs solve the problem of pose regulation with good robustness, they do not include a notion of memory, which allows the robot to recall the transverse path or to know where currently is with respect to a target. In order to reduce the dependence of a visual control scheme on the information in the image space and provide a notion of memory, we have dealt with the problem of pose-estimation and feedback of the estimated pose. In the literature, this has been tackled through static approaches by decomposition of the homography model or the essential matrix. In this context, the approach presented in this book is a novel nonlinear observability study, which demonstrates that the epipolar geometry and the trifocal tensor can be used for dynamic pose-estimation. This result allows the control of the robot in the Cartesian space with the benefit of facilitating the planning of complex tasks, like obstacle avoidance, in comparison to image-based approaches. We also demonstrate the stability of the closed loop with feedback of the estimated pose. Typically, position-based visual servoing schemes require a strict camera calibration process, however, the presented approach does not require specific calibration. Moreover, a target model, scene reconstruction or depth information are not needed. Additionally, the described control scheme is able to drive the robot to a desired position and also orientation through smooth velocities given by a single controller.

Before the details of the book, we want to thank to the Departamento de Informática e Ingeniería de Sistemas and the Instituto de Investigación en Ingeniería de Aragón from Universidad de Zaragoza, Spain, where the work was conducted, and also thanks to the Consejo Nacional de Ciencia y Tecnología (CONACYT), Mexico, for the support of the first author during his stay in Spain. Also, thanks to the CIMAT for the support on the publication of this work.

Guanajuato, Mexico, and Zaragoza, Spain
February 2014

Héctor M. Becerra
Carlos Sagüés

Contents

Chapter 1
Introduction

Abstract. In this chapter, we give a general introduction of the book, first specifying the context and the importance of the research on visual control of mobile robots. We have included a complete review of the state of the art on visual control. This review starts from a general point of view where the control is carried out in 6 degrees of freedom, and then, in the context of wheeled mobile robots. We also introduce the mathematical modeling of a wheeled mobile robot, the model of central cameras based on the generic camera model and the models of the visual measurements used for feedback, which are particularly taken from a multi-view geometric constraint.

1.1 Context of the Book

Mobile robots are the focus of much of the current research in the field of robotics. The importance of the research in mobile robots is still increasing, motivated by the introduction of this type of robots in service applications, where wheeled mobile robots are particularly appreciated. Among these applications we can include domestic and public cleaning, inspection and security patrols, transport of goods in hospitals, museum tour guides, entertainment with human interaction, and assistance to disable or elder people. Other typical applications of wheeled mobile robots are planetary exploration, exploration of inhospitable terrains, defusing explosives, mining and transportation in factories, ports and warehouses. Recently, there exist efforts in the development of autonomous personal transportation vehicles for places like airport terminals, attraction resorts and university campuses.

Although the autonomous navigation of robots is a mature field of research, it is still an open problem. In order to perform autonomous navigation, a robot must interact with its environment by using sensors. During a navigation task, a robot must answer three questions: *Where am I?*, *What the world looks like?* and *How to reach a desired place?* The first two questions are related to the perception and the last one is treated by the control system. Machine vision is a promising robotic sensor since the important advantage that provides the non-contact measurement of

H.M. Becerra and C. Sagüés, *Visual Control of Wheeled Mobile Robots*,
Springer Tracts in Advanced Robotics 103,
DOI: 10.1007/978-3-319-05783-5_1,

the environment. Moreover, this sensory modality is low price and it provides rich information from the environment, however, the optimal use of this information is still an appealing task.

Thus, visual navigation of mobile robots is a very interesting field of research from a scientific and even social point of view. Nowadays, the research efforts are focused on applications with monocular vision. On one hand, this presents special challenges as the lack of depth information because of a monocular imaging system is not a 3D sensor by itself. Additionally, the image processing for data interpretation is time consuming. However, given the great advances in computing power, the current impact of this aspect is less than in the past and will be eventually negligible in the close future. On the other hand, a monocular imaging system as a single sensor provides large range visibility as well as good precision in bearing measurements.

It is clear the importance of a control system in the context of robot navigation in order to answer the question *How to reach a desired place?* The control system must provide the suitable input velocities to drive the robot to the desired location. Additional to the challenges in the use of monocular vision as perception system, there exist some peculiarities in the control of wheeled mobile robots. Kinematically, a wheeled mobile robot is a nonlinear time-variant multivariable driftless system, which has non-holonomic motion constraints that results in an underactuated system. This causes a degree of freedom in the closed loop system that requires a particular control strategy to be driven to a desired value.

In general, when vision and control are joined in the so-called *visual servoing* approach, two main issues must be ensured, stability and robustness of the closed loop. Because of the nonlinearity of the problem, singularities frequently appear when the velocities are computed using an input-output transformation of the system. The control system must cope with these singularities in order to ensure the stability of the closed loop. There are many visual servoing schemes in the literature based on a pseudoinverse approach for non-square systems, which present potential problems of stability. Regarding to robustness, there exist parametric uncertainties due to calibration errors and measurement noise added to the feedback signals whose effects must be minimized.

There exist an important difference in how the desired value of the state of the system is given in a control loop using visual feedback in comparison with other type of feedback. Due to the lack of an absolute reference in the image space, the desired value of a visual control loop is set by giving a reference or target image. This implies that the target must be previously known and the measurements are relative values between the locations associated to the target image and the current image. It emphasizes the need of a learning phase in a visual control loop where the target must be memorized. Given that the target image encodes the desired robot pose, the control laws presented in this book deals with the problem of pose regulation, in which, simultaneous position and orientation correction is demanded.

In order to extract feedback information from the current and target images, both views require to share information, which means to have a common set of visual features in both images. In many cases, and especially when the initial camera position is far away from its desired value, the features belonging to the target may leave the camera field of view during servoing, which leads to failure because feedback error cannot be computed anymore. Therefore, particular schemes must be developed to cope with this problem and all the issues aforementioned.

1.2 State of the Art on Visual Control

Visual control is understood as the use of visual information as feedback in a closed loop control. This type of control approach represents a very good option for the control of robots, given the rich information that machine vision provides at a low cost. Additionally, the use of machine vision allows to be a small step closer to the goal of mimicking the human control action by an artificial robotic system. This section presents a review of the research on visual control, first from a general point of view, and then, in the context of wheeled mobile robots.

1.2.1 Visual Control in Robotics

The generic term visual control has been typically referred as *visual servoing*. Since the introduction of the concept in 1979 [66], an extensive work in the field has been developed. Visual servoing is currently a mature topic for robot control, although many aspects merit to be studied to make this approach more general and robust for conventional situations. It is generally accepted that visual servoing is defined as the use of visual information to control the pose of the robot's end-effector relative to a target object in the case of manipulators or the vehicle's pose with respect to some landmarks in the case of mobile robots [69].

The visual information needed for visual servoing may be acquired from a camera mounted on a robot manipulator or on a mobile robot (*eye-in-hand configuration*), in which case the robot motion induces camera motion. Also, the camera can be fixed in the workspace, so that, it can observe the robot motion from a stationary configuration (*eye-to-hand configuration*). Other configurations are possible, for instance, several cameras mounted on pan-tilt heads observing the robot motion. The mathematical development of all these cases is similar [31], and we focus on surveying the state of the art for the eye-in-hand case. This configuration increases the applicability of the visual servoing approach for the control of mobile robots given that it facilitates the carrying out of a large displacement task, which would be difficult using a fixed-camera configuration. The typical way in which a "setpoint" is given in visual control is through a reference or target image, which is referred as the *teach-by-showing* strategy. Thus, the goal of the visual servoing is to take the current view of the camera-robot to be the same as the target image by minimizing an error function $\mathbf{e}(t)$, which can be defined as

$$\mathbf{e}(t) = \mathbf{s}(\mathbf{m}(t), \mathbf{c}) - \mathbf{s}^d$$

where $\mathbf{s}$ is a vector of visual features depending on a set $\mathbf{m}(t)$ of visual measurements (e.g., the image coordinates of interest points or the image coordinates of the centroid of an object) and on a set **c** of metric parameters (e.g., camera calibration parameters or 3D information). The vector $\mathbf{s}^d$ contains the desired values of the features, which is constant in a static framework, i.e., fixed desired pose and motionless target. Moreover, $\mathbf{s}^d$ is time-varying if it is desired to drive the vector of visual features to a final value by tracking a particular trajectory.

The error function is often referred as a task function [123], [50], which must be zeroed $\mathbf{e}(t) = 0$. This can be seen as a problem of output regulation from a control theory viewpoint [70]. According to the nature of the error function, the visual servoing schemes are typically classified in three groups:

1. Image-based visual servoing (IBVS). The error is computed from a set of visual features **s** that are directly available in the image space.
2. Position-based visual servoing (PBVS). The error is computed in the Cartesian task space from a set of 3D parameters **s**, which must be estimated from visual measurements.
3. Hybrid visual servoing. The error function is a combination of Cartesian and image measurements.

Some interesting overviews introducing the different types of visual servoing schemes are [69] and the two-part survey [31], [32]. In Part I, classical visual servoing schemes are described and the performance of position-based and image-based visual servoing are analyzed. The Part II is dedicated to advance schemes, including partitioned, hybrid and numerical methods.

1.2.2 *Classical Visual Servoing Schemes*

Classical visual servoing schemes are based on the estimation of the so-called image Jacobian or interaction matrix, which relates the change of the robot velocities to the change of the visual features in the image space. In general, a visual servoing scheme relies on locating consistently a desired feature in each image of an input sequence, i.e., a spatio-temporal tracking process of visual cues. Relevant references on feature tracking algorithms for visual servoing purposes are [94] and [97].

1.2.2.1 Image-Based Visual Servoing Schemes

This approach, also called 2D visual servoing, is known to be robust against camera intrinsic parameters and many of the efforts of the research in visual servoing has been dedicated to this type of approach. However, classical image-based visual servoing presents problems of local minima and the convergence is constrained to a region around the desired pose. This is due to the need of stacking the image Jacobians of different image features to form the interaction matrix, which in general is a rectangular matrix that needs a pseudoinversion during the control execution. There exist a lot of examples of image-based visual servoing schemes in the literature. In one of the pioneer works [141], a hierarchical sensor-based control structure with the control level given by an adaptive image-based visual servoing scheme is introduced. An application of visual servoing for positioning a robot with respect to a moving object by tracking it and estimating its velocity is presented in [33]. A control theoretic formulation (linear time-invariant LTI) and a controller design method (linear quadratic LQ) for an image-based visual servoing with redundant features is proposed in [61]. In [44], motion in the image is used as input to the control system. This motion can be estimated without any a priori knowledge of the observed scene and thus, it does not need visual marks on the observed scene to retrieve geometric features. Some efforts have been developed to deal with the dependence on depth of the features in the classical image-based visual servoing. On this issue, an extended 2D visual servoing that augments the task function with an estimated depth distribution of the target points and an estimated camera model to avoid local minima is developed in [125]. Another approach dealing with the estimation of feature depth for image-based visual servoing using a nonlinear observer is presented in [90]. The image-based visual servoing schemes are sensitive to noise and often, they require to apply some filtering technique for smoothing the visual measurements. Lines and moments have been proposed to achieve robustness to image noise. Additionally, lines facilitate detection [7], and image moments remove potential problems with redundant points [30]. Image-based visual servoing has achieved good decoupling and linearizing properties for 6 degrees of freedom using image moments [131].

The image-based visual servoing approach has concerned about some important issues in order to overcome the drawbacks of the approach, for instance, to keep the features in the field of view or to avoid local minima [36]. In [128], the ability of a computer vision system to perceive the motion of an object in its field of view is addressed. A quantitative measure of motion is derived and called perceptibility, which relates the magnitude of the rate of change in an object's position to the magnitude of the rate of change in the image of that object. This has derived to path planning strategies for image-based visual servoing. In [98], the robot redundancy for dynamic sensor planning is used for visual servoing through the minimization of a secondary cost function. A partitioned approach for image-based visual servoing that overcomes the problem that the robot executes desired trajectories in the image, but which can be indirect and seemingly contorted in Cartesian space is introduced in [40]. A potential function that repels feature points from the boundary of the image plane is incorporated to guarantee that all features remain in the image. In [107], planning in image space and image-based visual servoing are coupled to solve the difficulties when the initial and desired robot positions are distant. The proposed method is based on the potential field approach. The work in [37] deals with the problem of visual servoing for robot manipulators taking into account constraints such as visibility, workspace and joint constraints, while minimizing a cost function such as spanned image area, trajectory length and curvature.

1.2.2.2 Position-Based Visual Servoing Schemes

The position-based visual servoing approach is also called 3D visual servoing because the control inputs are computed in the three dimensional Cartesian space. The pose of the target with respect to the camera is estimated from image features. To do that, a perfect geometric model of the target model and a calibrated camera are required. Results in computer vision about 3D reconstruction from two views have been applied successfully, making the knowledge of the object model unnecessary, however the calibration requirements cannot be avoided. The position-based visual control approach has been applied mainly for robot manipulators. Usually, a pose-estimation technique is used to recover the Cartesian required information. In [142], an extended Kalman filter-based estimation is used to design a control law in terms of relative positions between the end-effector and the target object. In [80], adaptive Kalman filtering techniques are explored to improve the precision of the estimation in position-based visual servoing for manipulators [127]. Two papers introduce visual servoing based on an observer-scheme for applications with manipulators [62], [64]. A visual tracking scheme that includes a kinematic model for the object to be tracked is used. To perform the visual servoing task, the authors of [103] use a nonlinear state feedback. They propose an exact model for parametrization of the pose that allows the control of the translation and rotation of the camera separately. This work attempts the common concern of visual control approaches to design decoupled schemes. Another concern of visual control is to keep the image features belonging to the target in the current view along the navigation. In [135], a position-based visual servoing approach that consists in tracking an iteratively computed trajectory is designed to guarantee that the target object remains in the field of view.

Currently, the research on visual control focuses on applications on monocular vision and the use of stereo vision has received less attention. An approach that strictly speaking is a position-based visual servoing scheme takes advantage of the 3D features provided by a stereo vision system [28]. 3D coordinates of any point observed in both images can be easily estimated by a simple triangulation process. Binocular vision has proved the benefit of robustness to calibration errors exploiting optical flow [58]. Alternative feature vectors combining 2D and 3D information from a stereo system have been proposed in [27]. That work shows that point depth and object pose produce an improved behavior in the control of the camera.

It is worth mentioning that both image-based and position-based visual servoing exhibit potential problems of stability [29], and only local stability can be theoretically demonstrated. In the case of image-based visual servoing it is due to the rectangular form (in general) of the interaction matrix that has a no trivial null space. In the case of position-based visual servoing any error in the estimation of the pose parameters required in the interaction matrix affects the stability property of the closed loop.

1.2.3 Visual Servoing through a Geometric Constraint

The classical visual control schemes use a batch of features directly extracted from the images to compute the error function. Another option to extract useful information in order to control a robot from monocular images is by means of geometric constraints that relate two or three views. Geometric constraints are imposed in the images when there are correspondences between features [60]. At present, three constraints are well known in computer vision and have been used for control purposes: the homography model, the epipolar geometry and the trifocal tensor. The use of geometric constraints has given origin to hybrid schemes that improve the performance of classical image-based visual servoing. In [48], the homography and the epipolar constraint are used to generate the optimal trajectory of the robot motion to reach the goal straightforwardly with decoupled translation and rotation. The 2-1/2D visual servoing which is based on the estimation of the partial camera displacement from the current to the desired camera poses at each iteration of the control law is proposed in [96]. This scheme does not need any geometric 3D model of the object as required in position-based visual servoing, and it ensures the convergence of the control law in the whole task space, unlike image-based visual servoing. An outstanding work in the hybrid visual control approach concerns about the stability analysis of a class of model-free visual servoing methods that can be hybrid or position-based visual servoing [95]. In any of both cases, these methods do not need a 3D model of the target object. Additionally, the visual servoing is decoupled by controlling the rotation of the camera separately from the rest of the system. In [34], a homography-based adaptive visual servo controller is developed to enable a robot end-effector to track a desired Euclidean trajectory as determined by a sequence of images. The error systems are constructed as a hybrid of pixels information and reconstructed Euclidean variables obtained by comparing the images and decomposing a homographic relationship.

A homography-based approach for image-based visual tracking and servoing is proposed in [23]. The visual tracking algorithm is based on an efficient second-order minimization method and its output is a homography linking the current and the reference image of a planar target. Using the homography, a task function isomorphic to the camera pose is designed, and thus, an image-based control law is proposed. In [35], an image-space path planner is proposed to generate a desired image trajectory based on a measurable image Jacobian-like matrix and an image-space navigation function. An adaptive homography-based visual servo tracking controller is then developed to navigate to a goal pose along the desired image-space trajectory while ensuring that the target points remain visible.

In order to avoid the largely over-constrained control commands resulting in monocular approaches, in [75], the authors propose to exploit explicitly the epipolar constraint of two stereo images. A hybrid switched-system visual servo method that utilizes both image-based and position-based visual servoing is presented in [54]. The switching strategy achieves stability in both the pose space and image space simultaneously. With this strategy is possible to specify neighborhoods for the image error and pose error that the state can never leave. In [13], the epipolar constraint is introduced for visual homing of robot manipulators. By using the epipolar geometry, most of the parameters (except depth) which specify the differences in position and orientation of the camera between the current and target images are recovered. The method is memoryless, in the sense that at every step the path to the target position is determined independently of the previous path. In [118], a visual servoing based on epipolar geometry for manipulators is reported. It discusses how the real value of the translation to be reached is unpredictable and proposes a control scheme which decouples the control of the rotation and the translation using the projective properties of the fundamental matrix. Additionally, The epipolar visual control has become in a powerful approach for mobile robots as will be described later.

The first work that proposes a robotic application of a trilinear constraint is [49]. The trifocal tensor has proved its effectiveness to recover the robot location in [59]. The onedimensional version of the trifocal tensor has been also proposed for hierarchical localization with omnidirectional images in [110]. Recently, the trifocal tensor has been introduced for visual servoing of robot manipulators in [126]. Schemes using higher order tensors with more than three views have been explored for visual odometry [38].

1.2.4 Robust Visual Servoing

Another important issue in visual servoing concerns for uncalibrated methods. Some good results toward the development of robust visual techniques used to guide robots in several real-word tasks are presented in [74]. The scheme attempts to keep the object in the field of view, or even centered, while the end-effector is approaching to it. A visual servoing scheme which is invariant to changes in camera-intrinsic parameters is presented in [93]. With this method, it is possible the camera positioning with respect to a nonplanar object given a reference image taken with a completely different camera. A dynamic quasi-Newton method for uncalibrated visual servoing that estimates the composite Jacobian at each step is used in [115]. An adaptive controller for image-based visual servoing using a fixed camera whose intrinsic and extrinsic parameters are unknown is presented in [81]. It proposes a depth-independent interaction matrix and estimates on-line the values of the unknown parameters. The results in [140] present an adaptive scheme that avoids the dependence of the scheme on depth information. It exploits the dynamics of manipulators to prove asymptotic convergence of the image errors. However, this approach needs additional feedback of the robot state (position and velocity). Additionally, the camera calibration parameters are estimated on-line using a bundle adjustment strategy that is sensitive to initial values.

Few efforts have been done to take some robust control techniques for vision-based feedback. One of this techniques that is well know for its property of robustness is the sliding mode control. This control technique provides the possibility to deal with parametric uncertainty due to weak camera calibration. On one hand, sliding mode control has been used for 3D trajectory tracking with a position-based visual servoing scheme through a pose reconstruction algorithm [144]. On the other hand, sliding mode control has been proposed for 6 degrees of freedom image-based visual servoing in [73], and recently combined with a pre-estimation of the robot state [78].

1.2.5 Omnidirectional Visual Servoing

One effective way to enhance the field of view of conventional cameras is to use mirrors in conjunction with lenses, i.e., catadioptric image formation systems.

The authors of [12] propose to use catadioptric cameras to observe the whole robot's articulated mechanism so that its joints can be tracked and controlled simultaneously. It has been shown that some results of stability and robustness obtained for conventional cameras are extendable to the entire class of central catadioptric systems [108]. For instance, some of these extensions are the 2-1/2D scheme of [2] and the path planning strategy proposed in [4].

The problem of controlling a robot from the projection of 3D straight lines in the image plane of central catadioptric systems is addressed in [5]. A generic interaction matrix for the projection of 3D straight lines is derived using the unifying imaging model [56]. A new decoupled image-based visual servoing scheme allowing the control of translational motion independently of rotation is proposed from the projection onto a unitary sphere in [132]. The scheme is based on moments that are invariant to 3D rotational motion. The homography model has been also exploited for omnidirectional visual servoing of 6 degrees of freedom. The authors of [22] propose an image-based control law that minimizes a task function isomorphic to the camera pose. They provide the theoretical proofs of the existence of the isomorphism and the local stability of the control law. Other geometric constraints have been exploited while using omnidirectional vision for visual servoing, particularly focused on the control of mobile robots, as detailed in the next section.

1.2.6 Visual Control of Mobile Robots

From the last decade, some Jacobian-based visual servoing schemes for mobile robots have been proposed as an extension of those schemes for 6 degrees of freedom, for instance [104], [136]. These works propose to add some degrees of freedom to the mobile platform by acting the camera to overcome the control issues related to the non-holonomic constraint of wheeled robots. This is avoided in [63], where two visual cues attached on the environment are controlled through an image-based visual servoing scheme that linearizes the dynamics of the non-holonomic cart transformed into the image plane. These image-based visual servoing schemes are sensitive to noise and wrong corresponding point features, and often, it is required to apply some robust filtering technique for smoothing and matching visual measurements [39]. Redundancy resolution methods have been employed for image-based visual servoing of non-holonomic mobile manipulators [89]. An strategy for visual motion planning in the image space is presented in [145].

Most of the visual servoing schemes have the drawback that the target may leave the camera field of view during the servoing, which leads to failure because the feedback error cannot be computed anymore. Some position-based visual servoing schemes have been proposed to avoid this problem. For instance, the switching controller of [53] uses an Euclidean homography-based reconstruction to estimate the robot pose. The parking problem with limited field of view is tackled in [111] with a switching control law, whose stability is analyzed through the theory of hybrid systems and the robot pose is estimated in closed form from the motion of the image features.

The position-based visual servoing approach has allowed to treat other problems, like robot cooperation [71], wall following, leader following and position regulation [45], or visual servoing for large displacement [52]. These schemes introduce a Kalman filtering approach to estimate features position of a known pattern scene, or to match a set of landmarks to a priori map and to estimate the robot pose from these visual observations.

A robust way to relate corresponding features from two views is by means of a geometric constraint [60]. Three constraints have been mainly employed in the context of visual servoing: the homography model, the epipolar geometry and the trifocal tensor. The homography model has been used on visual servoing for mobile robots in the form of a hybrid scheme [51]. This strategy uses the rotation and the scaled translation between the current and the target views, which implies to decompose the homography matrix and requires perfect knowledge of the camera calibration parameters. Similarly, the homography is decomposed to estimate the relative position and orientation of a car with respect to a leading vehicle for car platooning applications in [24]. The idea of using the elements of the homography matrix directly as feedback information is introduced in [120] for visual correction in homing tasks. Also, the homography decomposition has been partially avoided in [86]. However, the rotation is still required and it is obtained from the parallax of the image planes. This scheme has been extended to deal with field of view constraints while the robot is driven through optimal paths to the target in [82].

Regarding to the use of the epipolar geometry for visual servoing of wheeled mobile robots, the authors of [114] introduce a strategy based on the autoepipolar condition, a special configuration for the epipoles which occurs when the desired and current views undergo a pure translation. Although the epipolar geometry is a more general constraint than the homography model, its application for visual servoing results challenging due to its ill-conditioning for planar scenes, the degeneracy with short baseline and singularity problems for system control. The first issue has been solved by using a generic scene and the second has been tackled by commuting to a feature-based strategy [99] or homography-based control [84]. The singularity problems in epipolar-based control appear when the interaction matrix between the robot velocities and the rate of change of the epipoles becomes singular for some state of the robot. The approach in [99] takes into account the non-holonomic nature of a wheeled robot by driving one dimension of the epipoles to zero in a smooth way. However, in order to avoid the singularity, the motion strategy steers the robot away from the target while the lateral error is corrected, and after that, the robot moves backward to the target position. A more intuitive way to drive the robot directly toward the target has been addressed in [87], but the singularity is not treated. A recent work presents a visual control for mobile robots based on the elements of a 2D trifocal tensor constrained to planar motion [85]. The interaction matrix of this scheme is a rectangular Jacobian that may induce problems of stability or local minima.

Omnidirectional vision has been exploited for visual control of mobile robots. The method in [137] proposes a switching control law with feedback of the bearing angle of features and range discrepancies between the current and target views through the so-called Improved Average Landmark Vector. The generic approach for central catadioptric imaging systems presented in [5] is also valid for mobile robots control. The control objective is formulated in the catadioptric image space and the control law is designed from a robot state vector expressed in the same space in order to follow 3D straight lines. The auto-epipolar condition has also been exploited in visual servoing with central catadioptric cameras in [100], where together with the pixel distances between the current and target image features, an image-based control law is designed for holonomic mobile robots. Recently, an approach for non-holonomic robots proposes to exploit multiple homographies from virtual planes covering a large range of the omnidirectional view in order to extract most of the relevant information [83].

1.3 Mathematical Modeling

In general, the design of a control law relies on the knowledge of a model of the system to be controlled, as well as a model of the outputs of the system that can be measured. In this section, we describe the mathematical model of a wheeled mobile robot, the generic camera model and the model of the visual measurements used in the visual servoing schemes developed in this book.

1.3.1 The Camera-Robot Model

Many wheeled robotic platforms can be characterized by having a differential-drive motion capability. In the context of the book, this type of robots are controlled using visual feedback under the framework that is depicted in Fig. 1.1. A camera is fixed to the robot and eventually it is translated a known distance ℓ along the longitudinal axis $\mathbf{y}_R$. On one hand, the kinematic behavior of the robot $\{R\}$ with respect to a world frame $\{W\}$ can be expressed using the unicycle model

$$\begin{bmatrix} \dot{x} \\ \dot{y} \\ \dot{\phi} \end{bmatrix} = \begin{bmatrix} -\sin\phi & 0 \\ \cos\phi & 0 \\ 0 & 1 \end{bmatrix} \begin{bmatrix} \upsilon \\ \omega \end{bmatrix}. \tag{1.1}$$

On the other hand, the kinematic behavior of the onboard camera as induced by the robot motion, i.e., the kinematics of the point $\mathbf{c}^R = \begin{bmatrix} 0 & \ell \end{bmatrix}^T$ can be found using a general transformation between frames. Thus, the onboard camera motion is modeled by the following continuous time system

$$\begin{bmatrix} \dot{x} \\ \dot{y} \\ \dot{\phi} \end{bmatrix} = \begin{bmatrix} -\sin\phi & -\ell\cos\phi \\ \cos\phi & -\ell\sin\phi \\ 0 & 1 \end{bmatrix} \begin{bmatrix} \upsilon \\ \omega \end{bmatrix}. \tag{1.2}$$

This is a driftless affine system that can be written through the state vector $\mathbf{x} = \begin{bmatrix} x & y & \phi \end{bmatrix}^T$, input vector $\mathbf{u} = \begin{bmatrix} \upsilon & \omega \end{bmatrix}^T$, two input vector fields $g_1(\mathbf{x}) =$

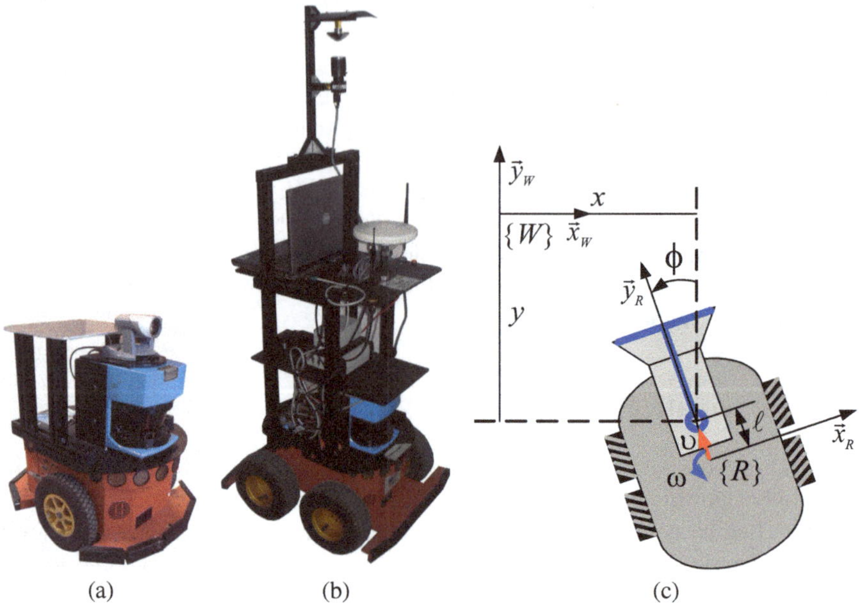

Fig. 1.1 Robotic test beds and robot model. (a) Non-holonomic mobile robot Pioneer P3-DX with a conventional camera onboard. (b) Non-holonomic mobile robot Pioneer 3-AT with a central catadioptric imaging system onboard. (c) Kinematic configuration of a mobile robot with an on-board central camera.

$\begin{bmatrix} -\sin\phi & \cos\phi & 0 \end{bmatrix}^T$ and $g_2(\mathbf{x}) = \begin{bmatrix} -\ell\cos\phi & -\ell\sin\phi & 1 \end{bmatrix}^T$, and output measurement $\mathbf{y}$ given by an adequate nonlinear function $h(\mathbf{x})$

$$\dot{\mathbf{x}} = \begin{bmatrix} g_1(\mathbf{x}) & g_2(\mathbf{x}) \end{bmatrix} \mathbf{u},$$
$$\mathbf{y} = h(\mathbf{x}). \tag{1.3}$$

The driftless property of this system is because the state vector field is null, and hence, no motion takes place under zero input, or in control theory concepts, any state is an equilibrium point under zero input. Furthermore, the corresponding linear approximation in any point $\mathbf{x}(t)$ is uncontrollable. However, it fulfills the Lie Algebra rank condition [70], in such a way that controllability can be demonstrated [91].

Additionally, the non-holonomic system (1.2) fails to satisfy a condition for stabilization via smooth time-invariant feedback (Brockett's theorem [25]). This theorem states that a necessary condition for smooth stabilization of a driftless regular system is that the number of inputs equal the number of states. Since this is not the case, such condition is violated. Thus, in order to drive the robot to a desired position and orientation (pose regulation problem), time-varying feedback control laws have been proposed in the literature [122], [116]. Also, this control problem has been tackled by maneuvering in a sequence of steps generating discontinuous control inputs [47], [6]. In all these control schemes, the full state of the robot is considered to be available for feedback. Particularly in the control of mobile robots using visual feedback, the authors of [51] and [136] propose time-varying controllers, but the maneuvering strategy has been mostly used [63], [39], [99], [87], [111], [82]. This strategy has been exploited for the design of control schemes considering the constrained field of view of conventional cameras, like in the last two referred works.

By applying an Euler approximation (forward difference) for the continuous derivatives in (1.2), the discrete version of the system is obtained

$$\begin{aligned} x_{k+1} &= x_k - T_s\left(\omega_k \ell \cos\phi_k + \upsilon_k \sin\phi_k\right), \\ y_{k+1} &= y_k - T_s\left(\omega_k \ell \sin\phi_k - \upsilon_k \cos\phi_k\right), \\ \phi_{k+1} &= \phi_k + T_s \omega_k, \end{aligned} \tag{1.4}$$

where T_s is the sampling time. Eventually, it can be assumed that the robot state and the measurements are affected by Gaussian noises $\mathbf{m}_k$ and $\mathbf{n}_k$, respectively. These noises accomplish $\mathbf{m}_k \sim N(0, \mathbf{M}_k)$, $\mathbf{n}_k \sim N(0, \mathbf{N}_k)$ and $E\left[\mathrm{m}_{k,i}\mathrm{n}_{k,j}^T\right] = 0$, with $\mathbf{M}_k$ the state noise covariance and $\mathbf{N}_k$ the measurement noise covariance. By writing the state vector as $\mathbf{x}_k = \left[x_k \; y_k \; \phi_k\right]^T$, the input vector as $\mathbf{u}_k = \left[\upsilon_k \; \omega_k\right]^T$, the discrete version of the system can be expressed as follows:

$$\begin{aligned} \mathbf{x}_{k+1} &= f\left(\mathbf{x}_k, \mathbf{u}_k\right) + \mathbf{m}_k, \\ \mathbf{y}_k &= h\left(\mathbf{x}_k\right) + \mathbf{n}_k, \end{aligned} \tag{1.5}$$

where the nonlinear function f is the vector field formed by the terms on the right side of (1.4).

1.3.2 Central Camera Model

One of the key point of this book is the development of visual control schemes valid for any type of camera obeying the so-called generic camera model described in this section. This increases significatively the applicability of the control schemes presented in the next chapters. The constrained field of view of conventional cameras (Fig. 1.2(a)) can be enhanced using wide field of view imaging systems such as full view omnidirectional cameras. This can be achieved using some optic arrangements that combine mirrors and lens, i.e., catadioptric imaging systems as the one in Fig. 1.2(b). These systems use hyperboloidal, paraboloidal or ellipsoidal mirrors and have been well studied in the field of computer vision [9]. According to this theory, all of them can satisfy the fixed view point constraint. In practice, with a careful construction of the system, it is realistic to assume a central configuration and many robotic applications have proven its effectiveness [106], [3], [100], [59].

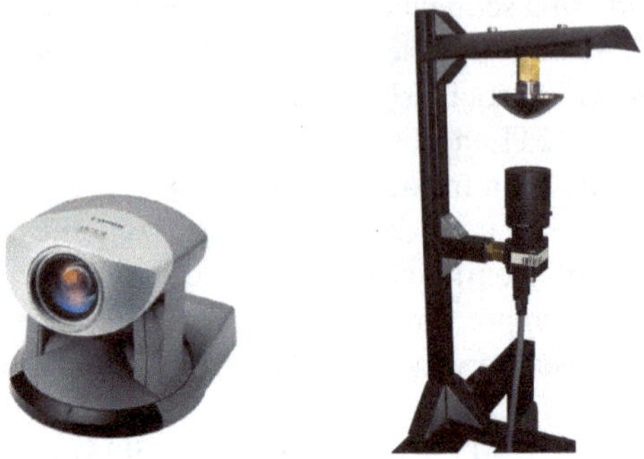

(a) (b)

Fig. 1.2 Examples of central cameras. (a) Conventional perspective camera. (b) Catadioptric imaging system formed by a hyperbolic mirror and a perspective camera.

It is known that the imaging process performed by conventional and catadioptric cameras can be modeled by a unique representation [56]. Such unified projection model works properly for imaging systems having a single center of projection. Although fisheye cameras do not accomplish such property, some recent experimental results have shown that the unified projection model is able to represent their image formation process with the required accuracy for robotic applications [41].

The unified projection model describes the image formation as a composition of two central projections [56]. The first is a central projection of a 3D point onto a virtual unitary sphere and the second is a perspective projection onto the image plane. According to [11], this generic model can be parameterized by (ξ,η) which are parameters describing the type of imaging system and by the matrix containing the intrinsic parameters. The parameter ξ encodes the nonlinearities of the image formation in the range $\xi \leq 1$ for catadioptric vision systems and $\xi > 1$ for fisheye cameras. The parameter η can be seen as a zooming factor and it is already included in the estimated value of the focal length. Thus, the parameter ξ and the generalized camera projection matrix $\mathbf{K}$ can be obtained through a calibration process using an algorithm for central catadioptric cameras like the one in [105]. This matrix is given as

$$\mathbf{K} = \begin{bmatrix} \alpha_x \eta & \alpha_x \eta s & x_0 \\ 0 & \alpha_y \eta & y_0 \\ 0 & 0 & 1 \end{bmatrix}, \tag{1.6}$$

where α_x and α_y represent the focal length of the camera in terms of pixel dimensions in the x and y direction respectively, s is the skew parameter and (x_0, y_0) are the coordinates of the principal point.

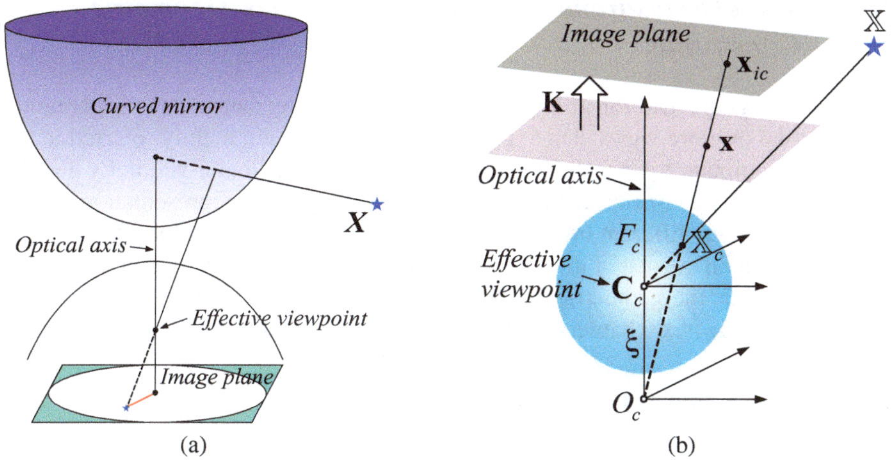

Fig. 1.3 The central image formation process. (a) Catadioptric imaging system. (b) Generic representation of central cameras.

Regarding to Fig. 1.3, the mapping of a point $\mathbb{X}$ in the 3D world with coordinates $\mathbf{X} = \begin{bmatrix} X & Y & Z \end{bmatrix}^T$ in the camera frame F_c resulting in the image point $\mathbf{x}_{ic}$ with homogeneous coordinates $\mathbf{x}_{ic}^h$ can be divided into three steps:

1. The world point is projected onto the unitary sphere on a point $\mathbb{X}_c$ with coordinates $\mathbf{X}_c$ in F_c, which are computed as $\mathbf{X}_c = \mathbf{X}/\left\|\mathbf{X}\right\|$.
2. The point coordinates $\mathbf{X}_c$ are then changed to a new reference frame O_c centered in $O = \begin{bmatrix} 0 & 0 & -\xi \end{bmatrix}^T$ and perspectively projected onto the normalized image plane $Z = 1-\xi$:

$$\mathbf{x}^h = \begin{bmatrix} \mathbf{x}^T & 1 \end{bmatrix}^T = \begin{bmatrix} x & y & 1 \end{bmatrix}^T = f(\mathbf{X})$$
$$= \begin{bmatrix} \frac{X}{Z+\xi\|\mathbf{X}\|} & \frac{Y}{Z+\xi\|\mathbf{X}\|} & 1 \end{bmatrix}^T.$$

3. The image coordinates expressed in pixels are obtained after a collineation $\mathbf{K}$ of the 2D projected point $\mathbf{x}_{ic}^h = \mathbf{K}\mathbf{x}^h$.

Notice that, setting $\xi = 0$, the general projection model becomes the well known perspective projection model. Images also depends on the extrinsic parameters $\mathbf{C} = (x, y, \phi)$, i.e. the camera pose in the plane relative to a global reference frame. Then an image is denoted by $I(\mathbf{K}, \mathbf{C})$.

It is possible to compute the coordinates of the point on the unitary sphere $\mathbf{X}_c$ from point coordinates on the normalized image plane $\mathbf{x}$ if the calibration parameters of the imaging system are know. As deduce in [42], the following holds:

$$\mathbf{X}_c = \left(\eta^{-1} + \xi\right)\bar{\mathbf{x}}, \tag{1.7}$$

where $\bar{\mathbf{x}} = \left[\mathbf{x}^T, \frac{1}{1+\xi\eta}\right]^T$ and $\eta = \frac{-\gamma-\xi\left(x^2+y^2\right)}{\xi^2\left(x^2+y^2\right)-1}$, $\gamma = \sqrt{1+\left(1-\xi^2\right)\left(x^2+y^2\right)}$.

1.3.3 *Visual Measurement's Models: Multi-view Geometric Constraints*

A multi-view geometric constraint is a mathematical entity that relates the geometry between two or more views. Three geometric constraints are mainly referred in the context of computer vision: the homography model, the epipolar geometry and the trifocal tensor. The last two constraints provide more general representations of the geometry between different images, since the former is valid only for planar views. For this reason, in the rest of this chapter, we describe the epipolar geometry and the trifocal tensor as the geometric constraints that have been exploited in the control approaches of the next chapters. For further details about this topic refer to [60].

1.3.3.1 The Epipolar Geometry

The epipolar geometry describes the intrinsic geometry between two views. It only depends on the relative location between cameras and their internal parameters. The fundamental matrix $\mathbf{F}$ is the algebraic representation of this geometry. This is a 3×3 matrix of rank 2. The fundamental matrix satisfies the epipolar constraint

$$\mathbf{p}_2^T \mathbf{F} \mathbf{p}_1 = 0,$$

where $\mathbf{p}_1$ and $\mathbf{p}_2$ is a pair of corresponding image points expressed in homogeneous coordinates. Using this constraint, the fundamental matrix can be computed from corresponding points in conventional images without knowledge of the internal camera parameters or the relative camera positions by solving a linear system of equations using the 8-point algorithm [60]. The epipoles $\mathbf{e}_1$ and $\mathbf{e}_2$ are the intersections of the line joining the optical centers of the cameras with the image planes. They can be computed from the fundamental matrix using the expressions $\mathbf{F}\mathbf{e}_1 = 0$ and $\mathbf{F}^T\mathbf{e}_2 = 0$.

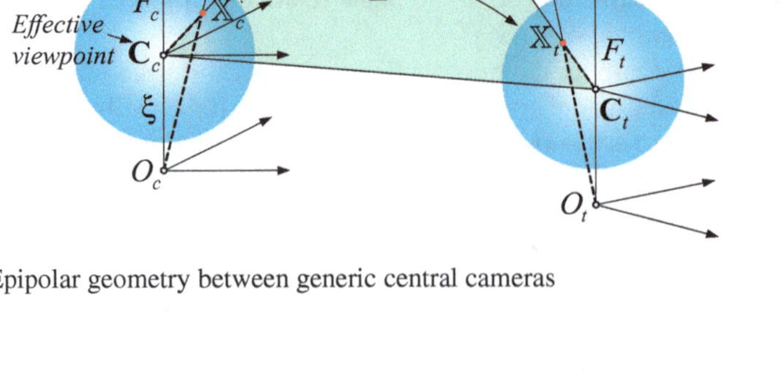

Fig. 1.4 Epipolar geometry between generic central cameras

The fundamental epipolar constraint is analogue for conventional cameras that for central catadioptric ones if it is formulated in terms of rays emanating from the effective viewpoint [130]. Regarding to Fig. 1.4, let $\mathbb{X}$ be a 3D point and let $\mathbf{X}_c$ and $\mathbf{X}_t$ be the coordinates of that point projected onto the unitary sphere of the current F_c and target frame F_t. These coordinates can be computed from the corresponding points in the images using (1.7). The epipolar plane contains the effective viewpoints of the imaging systems $\mathbf{C}_c$ and $\mathbf{C}_t$, the 3D point $\mathbb{X}$ and the points $\mathbb{X}_c$ and $\mathbb{X}_t$. The coplanarity of these points leads to the epipolar constraint for normalized cameras

$$\mathbf{X}_c^T \ \mathbf{E} \ \mathbf{X}_t = 0,$$

with $\mathbf{E}$ being the essential matrix. Normalized means that the effect of the known calibration matrix has been removed and then, central cameras can be virtually represented as conventional ones. As typical, from this constraint it is possible to compute the epipoles as the points lying on the baseline and intersecting the corresponding virtual image plane. This can be done by finding the right null space of the essential matrix. It is known that the estimation of the epipolar geometry degenerates with short baseline and becomes ill-conditioned for planar scenes [60].

Now, consider we have two images, a current $I_c(\mathbf{K}, \mathbf{C}_c)$ and a target one $I_t(\mathbf{K}, \mathbf{C}_t)$. Assuming a planar framework, these views have associated locations $\mathbf{C}_c = (x_c, y_c, \phi_c)$ and $\mathbf{C}_t = (x_t, y_t, \phi_t)$, respectively. Fig. 1.5(a) shows the epipoles in 3D of a configuration of the pair of virtual perspective cameras. Fig. 1.5(b) presents an upper view of this configuration, where the framework of the epipolar geometry constrained to planar motion is defined under a common reference frame.

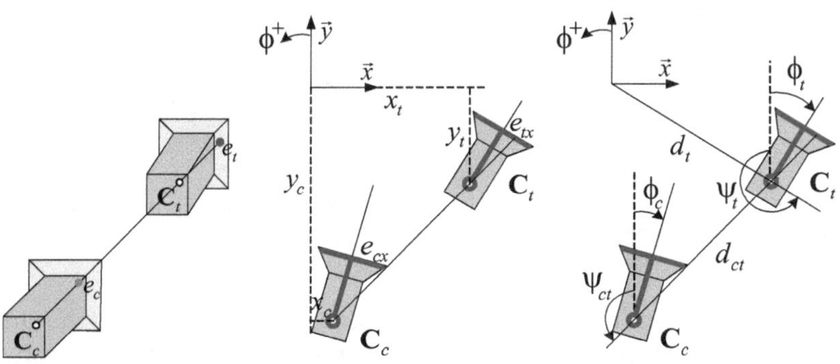

Fig. 1.5 Framework of the epipolar geometry. (a) 3D visualization of the epipolar geometry. (b) Epipoles from two views in the plane (e_{cx}, e_{tx}) and absolute positions with respect to a fixed reference frame. (c) Polar representation with relative parameters between cameras (d_{ct}, ψ_{ct}).

Fig. 1.5(c) depicts the locations in terms of their polar coordinates. In this general framework, the x-coordinate of the epipoles e_{cx} and e_{tx} can be expressed in terms of the camera locations and one calibration parameter as follows:

$$e_{cx} = \alpha_x \frac{(x_c - x_t)\cos\phi_c + (y_c - y_t)\sin\phi_c}{(y_c - y_t)\cos\phi_c - (x_c - x_t)\sin\phi_c}, \quad (1.8)$$
$$e_{tx} = \alpha_x \frac{(x_c - x_t)\cos\phi_t + (y_c - y_t)\sin\phi_t}{(y_c - y_t)\cos\phi_t - (x_c - x_t)\sin\phi_t}.$$

1.3.3.2 The Trifocal Tensor

The trifocal tensor plays a role in the analysis of scenes from three views analogous to the role played by the fundamental matrix in the two-view case. Like the fundamental matrix, the trifocal tensor depends nonlinearly on the motion parameters among the views, it encapsulates all the projective geometric relations between three views that are independent of the structure of the scene. The trifocal tensor arose historically as a relationship between corresponding lines in three views, however, it may be used to transfer points from a correspondence in two views to the corresponding point in a third view. The tensor can be computed from image correspondences alone without requiring knowledge of the motion or calibration.

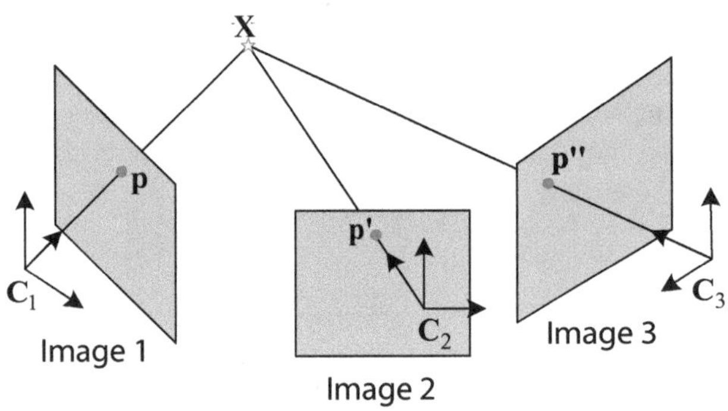

Fig. 1.6 3-point correspondences between points $\mathbf{p}$, $\mathbf{p}'$ and $\mathbf{p}''$ define the incidence correspondence through the trifocal tensor

The trifocal tensor consists of three 3×3 matrices ($\mathbf{T}_1$, $\mathbf{T}_2$, $\mathbf{T}_3$), and thus has 27 elements. There are therefore 26 independent ratios apart from the common overall scaling of the matrices. In this work we focus on the use of points as image features. Consider the corresponding points $\mathbf{p}$, $\mathbf{p}'$ and $\mathbf{p}''$ shown in Fig. 1.6 expressed in homogeneous coordinates. The incidence relation between those points is given as

$$[\mathbf{p}']_\times \left(\sum_i p^i \mathbf{T}_i\right) [\mathbf{p}'']_\times = \mathbf{0}_{3\times3},$$

where $[\mathbf{p}]_\times$ is the common skew symmetric matrix. This expression provides a set of nine equations, however, only four of them are linearly independent. It can be found that the four different choices of $i,l = 1,2$ give four different equations in terms of the observed image coordinates from the expression

$$p^k \left(p'^i p''^l T_k^{33} - p''^l T_k^{i3} - p'^i T_k^{3l} + T_k^{il} \right) = 0$$

for $k = 1,2,3$. Thus, seven triplets of point correspondences are needed to compute the elements of the tensor. The set of equations are of the form, $\mathbf{At} = \mathbf{0}$, where $\mathbf{A}$ is the equation matrix and $\mathbf{t}$ is a vector containing the elements of T_i^{jk} to be found. The solution to this problem constrained to $\|\mathbf{t}\| = 1$ in order to discard the solution $\mathbf{t} = \mathbf{0}$, can be found as the unit eigenvector corresponding to the least eigenvalue of $\mathbf{A}^T\mathbf{A}$. A good way to find this eigenvector is by using the singular value decomposition. The same as with the epipolar geometry, the estimation of the trifocal tensor becomes degenerated with short baseline, i.e., when the three camera locations are closer or are the same. However, the estimation of the tensor is well conditioned when the positions of the three cameras are collinear.

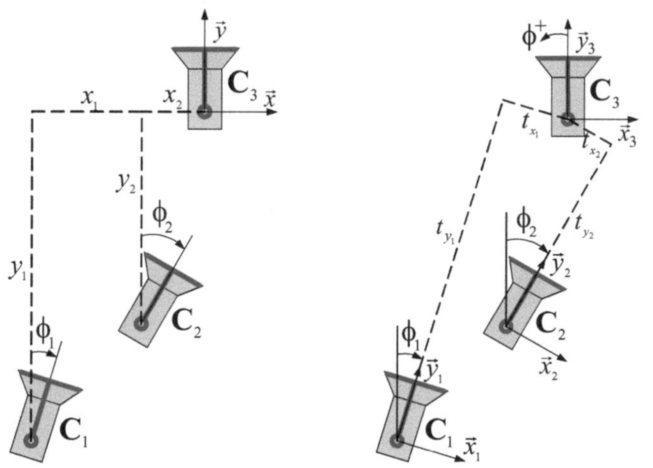

Fig. 1.7 Geometry between three camera locations in the plane. (a) Absolute locations with respect to a reference frame in $\mathbf{C}_3$. (b) Relative locations.

In the case in which the three cameras are located in the same plane, for instance, with the same vertical position from the ground, several elements of the tensor are zero and only 12 elements are in general non-null. Fig. 1.7 depicts the upper view of three cameras with global reference frame in the third view, in such a way that the corresponding locations are $\mathbf{C}_1 = (x_1, y_1, \phi_1)$, $\mathbf{C}_2 = (x_2, y_2, \phi_2)$ and $\mathbf{C}_3 = (0,0,0)$. Analytically, the trifocal tensor can be deduced for this framework as done in [85], resulting in that the non-null elements are given as

$$
\begin{aligned}
T^m_{111} &= -t_{x_1}\cos\phi_2 + t_{x_2}\cos\phi_1, \quad T^m_{113} = t_{x_1}\sin\phi_2 + t_{y_2}\cos\phi_1, \\
T^m_{131} &= -t_{y_1}\cos\phi_2 - t_{x_2}\sin\phi_1, \quad T^m_{133} = t_{y_1}\sin\phi_2 - t_{y_2}\sin\phi_1, \\
T^m_{212} &= -t_{x_1}, \; T^m_{221} = t_{x_2}, \; T^m_{223} = t_{y_2}, \; T^m_{232} = -t_{y_1}, \\
T^m_{311} &= -t_{x_1}\sin\phi_2 + t_{x_2}\sin\phi_1, \quad T^m_{313} = -t_{x_1}\cos\phi_2 + t_{y_2}\sin\phi_1, \\
T^m_{331} &= -t_{y_1}\sin\phi_2 + t_{x_2}\cos\phi_1, \quad T^m_{333} = -t_{y_1}\cos\phi_2 + t_{y_2}\cos\phi_1,
\end{aligned} \tag{1.9}
$$

where $t_{x_i} = -x_i\cos\phi_i - y_i\sin\phi_i$, $t_{y_i} = x_i\sin\phi_i - y_i\cos\phi_i$ for $i = 1,2$ and the superscript m states that they are the tensor elements given by metric information. In practice, the estimated tensor has an unknown scale factor and this factor changes as the robot moves. We can fix a common scale during the navigation by normalizing each element of the tensor as follows:

$$
\mathbf{T}_{ijk} = \frac{\mathbf{T}^e_{ijk}}{T_N}, \tag{1.10}
$$

where $\mathbf{T}^e_{ijk}$ are the estimated trifocal tensor elements obtained from point matches, $\mathbf{T}_{ijk}$ are the normalized elements and T_N is a suitable normalizing factor, which must be different than zero. We can see from (1.9) that T_{212} and T_{232} are constant and non-null, assuming that the camera location $\mathbf{C}_1$ is different to $\mathbf{C}_3$. Therefore, these elements are good option as normalizing factors.

Chapter 2
Robust Visual Control Based on the Epipolar Geometry

Abstract. In this chapter, we present a new control scheme that exploits the epipolar geometry but, unlike previous approaches based on two views, it is extended to three views, gaining robustness in perception. Additionally, robustness is also improved by using a control law based on sliding mode theory in order to solve the problem of pose regulation. The core of the chapter is a novel control law that achieves total correction of the robot pose with no auxiliary images and no 3D scene information, without need of commuting to any visual information other than the epipolar geometry and applicable with any central camera. Additionally, the use of sliding mode control avoids the need of a precise camera calibration in the case of conventional cameras and the control law deals with singularities induced by the epipolar geometry. The effectiveness of the approach is tested via simulations, with kinematic and dynamic models of the robot, and real-world experiments.

2.1 Introduction

This chapter presents an approach to drive a wheeled mobile robot to a desired location using the teach-by-showing strategy and exploiting the epipolar geometry in a robust image-based visual servoing scheme. The epipolar geometry describes the intrinsic geometry between two views and only depends on the relative location between cameras and their internal parameters [60], [113]. The epipolar geometry was introduced for control of robot manipulators in [13] and [118] around one decade ago.

This geometric constraint has shown some drawbacks, namely, the ill-conditioning for planar scenes, the degeneracy with short baseline and singularity problems for system's control. In related works about visual servoing of mobile robots, the first issue has been solved by using a generic scene and the second has been tackled by commuting to a feature-based strategy [99] or homography-based control [84]. The singularity problems in epipolar-based control appear when the interaction matrix between the robot velocities and the rate of change of the epipoles

H.M. Becerra and C. Sagüés, *Visual Control of Wheeled Mobile Robots*,
Springer Tracts in Advanced Robotics 103,
DOI: 10.1007/978-3-319-05783-5_2,

becomes singular for some state of the robot. Indeed, unbounded velocities eventually appear because the singularity is always reached when the robot moves directly toward the target. The approach in [99] takes into account the non-holonomic nature of a wheeled robot by driving one dimension of the epipoles to zero in a smooth way. However, in order to avoid the singularity, the motion strategy steers the robot away from the target while the lateral error is corrected, and after that, the robot moves backward to the target position. A more intuitive way to drive the robot directly toward the target has been addressed in [87] but the singularity is not treated. Another work that exploits the epipolar geometry, particularly the auto epipolar condition has been developed for holonomic mobile robots using central catadioptric cameras [100].

The method that we present in this chapter turns out in a control law based on sliding mode theory and feedback from the epipolar geometry in order to servo differential-drive mobile robots. The notion of this approach has been introduced in [17], where a robust two view-based control law is proposed, which is able to correct orientation and lateral error but not longitudinal error using conventional cameras. This scheme has been extended for central catadioptric cameras in [16]. Later, in [15], the epipolar geometry of three views is proposed to correct also longitudinal error only from the epipolar constraint.

As detailed in this chapter, the control strategy is performed in two steps to achieve position and orientation correction. In contrast to two-view based schemes, longitudinal error is also corrected by exploiting the epipolar geometry that relates a third image with the target image. This is done on the basis of a square control system, where global stability can be ensured. Additionally, this approach does not rely on any particular condition and takes into account the non-holonomic nature of a mobile platform. Our scheme does not need any geometric decomposition or additional parameter estimation to achieve pose regulation. The use of a third image allows to unify the control scheme in only one type of image-based controller for the whole task.

An important benefit of the scheme of this chapter with respect to previous epipolar-based approaches is that the control law corrects position and orientation by keeping full control during the whole task using only epipolar feedback. The control law copes with singularities induced by the epipolar geometry also improving the robot behavior by performing a direct motion toward the target. Besides, the use of the sliding mode control technique allows robust global stabilization of the task function (including image noise) when dealing with the weak calibration problem, i.e., no specific calibration is needed.

The rest of the chapter is organized as follows. Section 2.2 describes the pairwise epipolar geometry of three views. Section 2.3 details the design procedure of the sliding mode control law. Section 2.4 presents the stability and robustness analysis. Section 2.5 shows the performance of the closed-loop control system via simulations and real-world experiments, and finally, Section 2.6 summarizes the conclusions.

2.2 Pairwise Epipolar Geometry of Three Views

Although the epipolar geometry relates two views of an scene, the epipolar geometries of three views provides rich information that we exploit for visual servoing tasks. According to Fig. 2.1(a) and using the general framework for a pair of views described in section 1.3.3.1, the three pairings of epipolar relationships among three views can be found. Let us define a global reference frame with origin in the location of a third camera. Then, the camera locations with respect to that global reference are $\mathbf{C}_1 = (x_1, y_1, \phi_1)$, $\mathbf{C}_2 = (x_2, y_2, \phi_2)$ and $\mathbf{C}_3 = (x_3, y_3, \phi_3) = (0,0,0)$. Consider that such images have been taken by a camera mounted on a wheeled mobile robot where the camera reference frame coincides with the robot frame. In this camera-robot configuration, the x-coordinate of the epipoles can be written as a function of the robot state and the parameter α_x. The double subscript refers to the related images, for instance, e_{13} is the epipole in image one, as computed with respect to image three.

$$e_{13} = \alpha_x \frac{x_1 \cos\phi_1 + y_1 \sin\phi_1}{y_1 \cos\phi_1 - x_1 \sin\phi_1}, \quad e_{31} = \alpha_x \frac{x_1}{y_1}, \tag{2.1}$$

$$e_{23} = \alpha_x \frac{x_2 \cos\phi_2 + y_2 \sin\phi_2}{y_2 \cos\phi_2 - x_2 \sin\phi_2}, \quad e_{32} = \alpha_x \frac{x_2}{y_2}, \tag{2.2}$$

$$\begin{aligned} e_{12} &= \alpha_x \frac{(x_1 - x_2)\cos\phi_1 + (y_1 - y_2)\sin\phi_1}{(y_1 - y_2)\cos\phi_1 - (x_1 - x_2)\sin\phi_1}, \\ e_{21} &= \alpha_x \frac{(x_1 - x_2)\cos\phi_2 + (y_1 - y_2)\sin\phi_2}{(y_1 - y_2)\cos\phi_2 - (x_1 - x_2)\sin\phi_2}. \end{aligned} \tag{2.3}$$

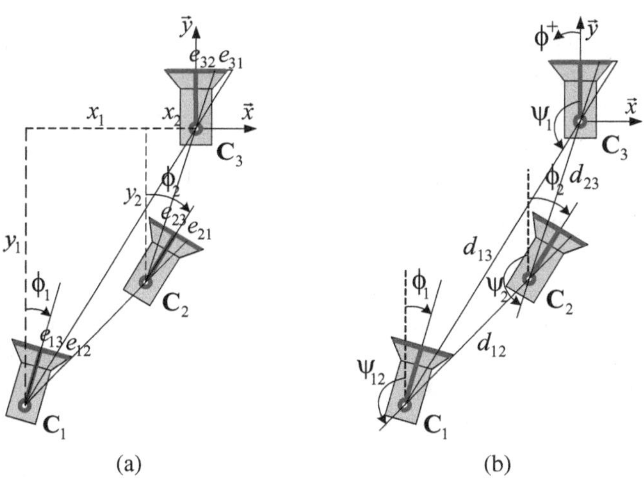

Fig. 2.1 Framework of the epipolar geometry for three views. (a) Epipoles from three views. (b) Polar coordinates.

The Cartesian coordinates for the camera location $\mathbf{C}_2$ can be expressed as a function of the polar coordinates d_{23} and ψ_2 (Fig. 2.1(b)) using

$$x_2 = -d_{23}\sin\psi_2,\ y_2 = d_{23}\cos\psi_2, \tag{2.4}$$

with $\psi_2 = -\arctan(e_{32}/\alpha_x)$, $\phi_2 - \psi_2 = \arctan(e_{23}/\alpha_x)$ and $d_{23}^2 = x_2^2 + y_2^2$.

The relative Cartesian coordinates between $\mathbf{C}_1$ and $\mathbf{C}_2$ can be expressed as a function of the polar coordinates d_{12} and ψ_{12} as follows:

$$(x_1 - x_2) = -d_{12}\sin\psi_{12},\ (y_1 - y_2) = d_{12}\cos\psi_{12}, \tag{2.5}$$

with $\psi_{12} = \phi_2 - \arctan(e_{21}/\alpha_x) = \phi_1 - \arctan(e_{12}/\alpha_x)$ and $d_{12}^2 = (x_1 - x_2)^2 + (y_1 - y_2)^2$.

Recall that all of these previous expressions are also valid for normalized cameras, i.e., computing the epipoles from the corresponding points on the unitary sphere as described in section 1.3.3.1. In such a case, the focal length parameter in the x-direction is $\alpha_x = 1$.

2.3 Epipolar Control Law from Three Views

Henceforth, let $\mathbf{C}_1$ be the initial camera location, $\mathbf{C}_2$ the current camera location and $\mathbf{C}_3$ the target camera location. The control strategy is performed in two steps as follows:

- First step - Alignment with the target: orientation and lateral error are corrected. This is achieved by zeroing the epipoles relating the current image $I_2(\mathbf{K}, \mathbf{C}_2(t))$ and the target one $I_3(\mathbf{K}, \mathbf{0})$. It can be seen as a two-view approach because only requires the epipoles e_{23} and e_{32}. Initially, we have two images (Fig. 2.2(a)) and at the end of this step, the robot is as in Fig. 2.2(b).
- Second step - Depth correction: pure translation along the $y-$axis. Provided that the orientation and the lateral error are maintained in zero by the control loop, the objective of this step is to achieve $e_{12} = e_{13}$ or $e_{21} = e_{31}$. This step requires the three images to compute the constant epipoles e_{13}, e_{31} from $I_1(\mathbf{K}, \mathbf{C}_1)$, $I_3(\mathbf{K}, \mathbf{0})$ and the varying epipoles e_{12}, e_{21} from $I_1(\mathbf{K}, \mathbf{C}_1)$, $I_2(\mathbf{K}, \mathbf{C}_2(t))$.

Finally, after the first step has corrected lateral error, the epipolar geometry in the three-view configuration is $e_{12} = e_{13}$, $e_{21} = e_{31}$, which implies $I_2(\mathbf{K}, \mathbf{C}_2) = I_3(\mathbf{K}, \mathbf{0})$, and consequently $\mathbf{C}_2 = \mathbf{C}_3$ as desired (Fig. 2.2(c)).

We assume that the robot is initially in a general configuration, not aligned with the target pose. Otherwise, this particular configuration can be trivially detected from the epipoles and, in that case, a simple initial motion controlling the epipoles can drive the robot to a general configuration.

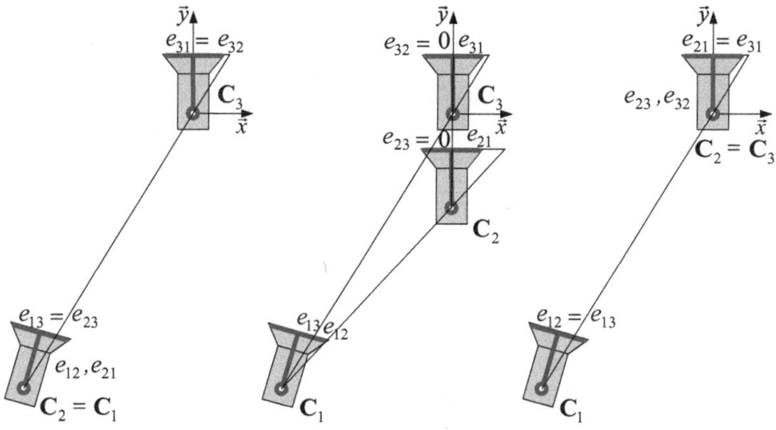

Fig. 2.2 Control strategy from three views. (a) Initial configuration. (b) Intermediate configuration. (c) Final configuration.

2.3.1 *First Step – Alignment with the Target*

The control objective of this step is zeroing the epipoles relating the current and target images. It means a simultaneous correction of orientation and lateral error. There exist two main drawbacks with the epipolar geometry in two views: uncertainty in parameters and singularity problems. This section describes the synthesis of a control law from two images, which copes with these both issues. The objective of this step is to perform the navigation toward the target by using the feedback information provided by the x-coordinate of the epipoles that relate the current image $I_2(\mathbf{K}, \mathbf{C}_2(t))$ and the target one $I_3(\mathbf{K}, \mathbf{0})$. The robot performs a smooth direct motion toward the target position applying adequate velocities during the whole task using the same robust control scheme even in singular situations.

Let us define the outputs of the system using the x-coordinates of the epipoles for the current and target images. Then, the two-dimensional output of the camera-robot system is

$$\mathbf{y} = h(\mathbf{x}) = \left[e_{23} \; e_{32} \right]^T . \tag{2.6}$$

In addition to the drawbacks of controlling a robot from two views, longitudinal error correction cannot be reached with only such information. The camera-robot system has relative degree-2 because the control inputs appear in the first time-derivative of the epipoles and the system is input-output linearizable with first order *zero dynamics* (see appendix A.1 and [70] for details). This unobservable dynamics is derived by making the epipoles defined as outputs equal to zero and then finding out the robot state. In the particular case of the camera-robot system (1.2) with $\ell = 0$ and output vector (2.6), this set, denoted by Z^*, is

$$Z^* = \{\mathbf{x} \mid e_{23} \equiv 0,\; e_{32} \equiv 0\} = \left\{ \left[0 \; y_2 \; 0 \right]^T ,\; y_2 \in \mathbb{R} \right\}. \tag{2.7}$$

The zero dynamics in this control system means that, when epipoles relating the moving view and the target one are zero, the x-coordinate of the robot position and the orientation are corrected, but the longitudinal error may be different to zero. As mentioned previously, this is corrected in a second step. Let us define tracking error functions as $\xi_c = e_{23} - e_{23}^d$ and $\xi_t = e_{32} - e_{32}^d$, where e_{23}^d and e_{32}^d are suitable time-varying references. From the time-derivatives of these errors and using the polar coordinates (2.4), we obtain the error system

$$\begin{bmatrix} \dot{\xi}_c \\ \dot{\xi}_t \end{bmatrix} = \begin{bmatrix} \frac{-\alpha_x \sin(\phi_2 - \psi_2)}{d_{23}\cos^2(\phi_2 - \psi_2)} & \frac{\alpha_x}{\cos^2(\phi_2 - \psi_2)} \\ \frac{-\alpha_x \sin(\phi_2 - \psi_2)}{d_{23}\cos^2(\psi_2)} & 0 \end{bmatrix} \begin{bmatrix} \upsilon \\ \omega \end{bmatrix} - \begin{bmatrix} \dot{e}_{23}^d \\ \dot{e}_{32}^d \end{bmatrix}. \quad (2.8)$$

The system (2.8) has the form $\dot{\xi} = \mathbf{M} \cdot \mathbf{u} - \dot{\mathbf{e}}^d$, where $\mathbf{M}$ corresponds to the decoupling matrix. We use the term decoupling matrix, coined in [129], instead of features Jacobian or interaction matrix in order to better describe the action of this matrix in the frame of control theory. The inverse of the matrix $\mathbf{M}$ is

$$\mathbf{M}^{-1} = \frac{1}{\alpha_x} \begin{bmatrix} 0 & -\frac{d_{23}\cos^2(\psi_2)}{\sin(\phi_2 - \psi_2)} \\ \cos^2(\phi_2 - \psi_2) & -\cos^2(\psi_2) \end{bmatrix}, \quad (2.9)$$

and $\dot{\mathbf{e}}^d$ represents a feedforward control term. In order to invert the system (2.8) applying the input-output linearization technique, it is important to notice that $\mathbf{M}$ loses rank if $\phi_2 - \psi_2 = n\pi$ with $n \in \mathbb{Z}$. This makes the element of the first row of (2.9) grow unbounded and, consequently, the translational velocity as well. As can be seen in the analytical expression of the inverse matrix (2.9), the computation of input velocities is bounded for any other situation. From the definition of the angles below (2.4), it can be seen that the singular condition corresponds to $e_{23} = 0$. This is a problem because it is indeed a control objective.

We faced the tracking problem as the stabilization of the error system (2.8), which has uncertain parameters α_x and d_{23}, i.e., focal length and distance between cameras. These parameters are estimated as the constants α_{x_e} and d_{23_e}, and introduced to the estimated decoupling matrix $\mathbf{M}_e(\phi_2, \psi_2)$.

We present a robust control law based on sliding mode control (see appendix A.2 and [138] for details) in order to tackle two main issues: 1) the sensitivity to uncertainty in parameters of a control system based on input-output linearization, which degenerates the performance of the tracking, see for example [87] and 2) the need to maintain control during the entire navigation even when the singularity occurs.

2.3.1.1 Decoupling-Based Control Law

First, let us define the following sliding surfaces:

$$\mathbf{s} = \begin{bmatrix} s_c \\ s_t \end{bmatrix} = \begin{bmatrix} \xi_c \\ \xi_t \end{bmatrix} = \begin{bmatrix} e_{23} - e_{23}^d \\ e_{32} - e_{32}^d \end{bmatrix} = \mathbf{0}. \quad (2.10)$$

Thus, the tracking problem is solved if there exist switched feedback gains according to an stability analysis that make the state to evolve on $\mathbf{s} = \mathbf{0}$. The following sliding mode control law, consisting of a so-called equivalent control (feedforward term) and a two-dimensional vector of switched feedback gains, ensures global stabilization of the system (2.8):

$$\mathbf{u}_{sm} = \mathbf{M}_e^{-1}(\phi_2, \psi_2) \begin{pmatrix} \dot{e}_{23}^d - \kappa_c \mathrm{sign}(s_c) \\ \dot{e}_{32}^d - \kappa_t \mathrm{sign}(s_t) \end{pmatrix}, \tag{2.11}$$

with $\kappa_c > 0$ and $\kappa_t > 0$ being control gains. The action of switched feedback gains in the error dynamics is to keep the state trajectory on the sliding surface (2.10). These gains add or subtract accordingly, in order to force the state trajectory to head always toward the surface [138], [67]. When the state trajectory crosses the surface because noise or drift, the control switches from addition to subtraction or viceversa in such a way that the trajectory reverses its direction and heads again toward the surface.

Although (2.11) can achieve global stabilization of (2.8), it may need high gains that could cause a non-smooth behavior in the robot state, which is not valid in real situations. Therefore, we add a pole placement term in the control law to alleviate this problem

$$\mathbf{u}_{pp} = \mathbf{M}_e^{-1}(\phi_2, \psi_2) \begin{bmatrix} -\lambda_c & 0 \\ 0 & -\lambda_t \end{bmatrix} \begin{bmatrix} s_c \\ s_t \end{bmatrix},$$

where $\lambda_c > 0$ and $\lambda_t > 0$ are control gains. Finally, the complete sliding mode control law ($\mathbf{u} = \mathbf{u}_{db}$) that achieves robust global stabilization of the system (2.8) is

$$\mathbf{u}_{db} = \begin{bmatrix} \upsilon_{db} \\ \omega_{db} \end{bmatrix} = \mathbf{u}_{sm} + \mathbf{u}_{pp}. \tag{2.12}$$

2.3.1.2 Bounded Control Law

The control law (2.12) utilizes the decoupling matrix that presents a singularity problem when the camera axis of the robot is aligned with the baseline, which generates unbounded translational velocity. In order to pass through the singularity we commute to a direct sliding mode controller when $|\phi_2 - \psi_2|$ is below a threshold T_h. This kind of controller has been studied for output tracking through singularities [67]. In this case, an adequate direct sliding mode controller is as follows:

$$\mathbf{u}_b = \begin{bmatrix} \upsilon_b \\ \omega_b \end{bmatrix} = \begin{bmatrix} -k_\upsilon \mathrm{sign}(s_t\, b(\phi_2, \psi_2)) \\ -k_\omega \mathrm{sign}(s_c) \end{bmatrix}, \tag{2.13}$$

where k_υ and k_ω are suitable gains and $b(\phi_2, \psi_2)$ is a function that describes the change in sign of the translational velocity when the state trajectory crosses the singularity. This function can be deduced from the first row of $\mathbf{M}^{-1}$ (2.9) as

$$b(\phi_2, \psi_2) = -\sin(\phi_2 - \psi_2). \tag{2.14}$$

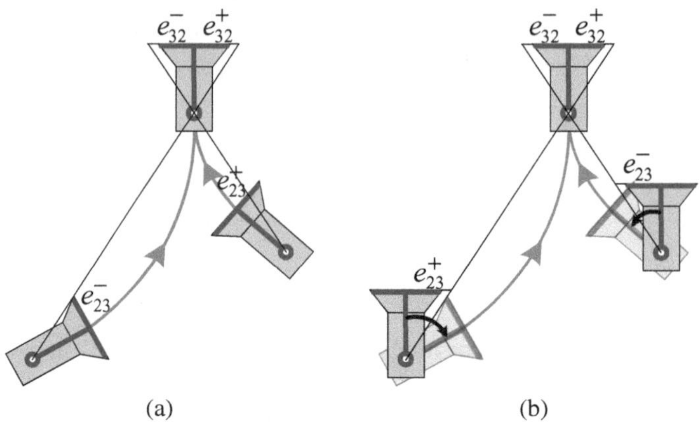

Fig. 2.3 Different cases for control initialization through the desired trajectories. (a) $\text{sign}(e_{32}) \neq \text{sign}(e_{23})$- direct motion toward the target. (b) $\text{sign}(e_{32}) = \text{sign}(e_{23})$ - rotation to reach the same condition as in (a).

The control law (2.13) with $b(\phi_2, \psi_2)$ (2.14) locally stabilizes the system (2.8) and is always bounded.

2.3.1.3 Desired Trajectories for the Epipoles

As main requirement, the desired trajectories must provide a smooth zeroing of the epipoles from their initial values. Fig. 2.3(a) shows two configurations of robot locations for cases in which $\text{sign}(e_{23}) \neq \text{sign}(e_{32})$. From these conditions, the epipoles are naturally reduced to zero as the robot moves directly toward the target. In order to carry out this kind of motion, the locations starting $\text{sign}(e_{23}) = \text{sign}(e_{32})$ need to be controlled to the situation of $\text{sign}(e_{23}) \neq \text{sign}(e_{32})$. In such a case, the control law forces the robot to rotate initially to reach an adequate orientation (Fig. 2.3(b)). It is worth emphasizing that this initial rotation is autonomously carried out through the control inputs given by the described controllers. The following trajectories provide the described behavior:

$$
\begin{aligned}
e_{23}^d(t) &= \sigma \frac{e_{23}(0)}{2}\left(1+\cos\left(\frac{\pi}{\tau}t\right)\right), & 0 \le t \le \tau, \\
e_{23}^d(t) &= 0, & \tau < t < \infty, \\
e_{32}^d(t) &= \frac{e_{32}(0)}{2}\left(1+\cos\left(\frac{\pi}{\tau}t\right)\right), & 0 \le t \le \tau, \\
e_{32}^d(t) &= 0, & \tau < t < \infty,
\end{aligned}
\tag{2.15}
$$

where $\sigma = -\text{sign}(e_{23}(0)e_{32}(0))$ and τ is the time to perform the first step of the control strategy. In our approach, as in any image-based scheme, the desired trajectories in the image space play an important role in the resultant Cartesian path.

By changing the reference trajectory of the target epipole (related to the translational velocity) is possible to run our approach for car-like robots.

Both previous controllers, i.e., (2.12) and (2.13), can be seen as a commuted control law and their stability is shown later. The control law is able to track the previous references using bounded velocities and its termination condition is set with the time τ.

2.3.2 Second Step – Depth Correction with Drift Compensation

The commuted controller of the first step is only able to correct orientation and lateral error due to the zero dynamics (2.7). We have described that a third image allows to define an appropriate control goal to correct the remaining depth. This third image is the initial one and it does not introduce an expensive computational load, given that the corresponding image points are already known and only the 8-point algorithm has to be run. This second step is treated as a regulation problem with integral action to give steady state robustness to the control loop [72], since we have constant desired values (e_{13}, e_{31}).

Let us define error functions $\xi_{12} = e_{12} - e_{13}$ and $\xi_{21} = e_{21} - e_{31}$. We can see from (2.3) that ξ_{12} does not depend on the rotation and, to avoid coupling problems between inputs, we have chosen its dynamics to work out the translational velocity. Let us define an augmented error system for ξ_{12}, whose dynamics is obtained using (2.5)

$$\begin{aligned} \dot{\xi}_{12}^{0} &= e_{12} - e_{13} = \xi_{12}, \\ \dot{\xi}_{12} &= \frac{\alpha_x \sin(\phi_2 - \psi_{12})}{d_{12} \cos^2(\phi_1 - \psi_{12})} \upsilon, \end{aligned} \tag{2.16}$$

where the new state ξ_{12}^{0} corresponds to the integral of the error. A common way to define a sliding surface is a linear combination of the state as follows:

$$s = k_0 \xi_{12}^{0} + \xi_{12} = 0, \tag{2.17}$$

in such a way that when $s = 0$ we have $\xi_{12} = -k_0 \xi_{12}^{0}$. By substituting this expression in (2.16), the reduced order system $\dot{\xi}_{12}^{0} = -k_0 \xi_{12}^{0}$ is obtained. It is clear that for any $k_0 > 0$ the reduced dynamic ξ_{12}^{0} is exponentially stable, and similarly ξ_{12}. We make $\dot{s} = 0$ to find the equivalent control, and then, a switching feedback gain is added to yield

$$\upsilon_{dc} = \frac{d_{12_e} \cos^2(\phi_1 - \psi_{12})}{\alpha_{x_e} \sin(\phi_2 - \psi_{12})} \left(-k_0 \xi_{12} - k_1 \mathrm{sign}(s) \right), \tag{2.18}$$

where $k_1 > 0$ is a control gain. Notice that $\sin(\phi_2 - \psi_{12})$ is never zero for the situation displayed in Fig. 2.2(b). This control law achieves robust global stabilization of the system (2.16) and its termination condition can be given by verifying that $e_{12} - e_{13} \approx 0$.

Although only a straight motion is needed during this second step, orientation control is maintained in order to compensate for the noise or drift that is always present in real situations. The bounded rotational velocity (2.13) is kept during the second step. However, this velocity depends on e_{23}, which has the problem of short baseline when the target is reached. In order to alleviate this issue, we use a similar approach to the one presented in [87]. An intermediate image is used instead of the target one when the epipolar geometry degenerates. In our case, the last current image of the first step is stored, which corresponds to an image acquired from a location aligned to the target. This intermediate image is denoted by $I_{2_{ar}}(\mathbf{K}, \mathbf{C}_{2_{ar}})$, where the subscript *ar* stands for "aligned robot". So, the computation of the rotational velocity as the complement of the translational velocity υ_{dc} (2.18) during the second step is carried out from the adequate images as follows:

$$\begin{aligned} I_2(\mathbf{K}, \mathbf{C}_2), I_3(\mathbf{K}, \mathbf{0}) &\implies \omega_b = -k_\omega \text{sign}(e_{23}), \\ I_{2_{ar}}(\mathbf{K}, \mathbf{C}_{2_{ar}}), I_2(\mathbf{K}, \mathbf{C}_2) &\implies \omega_b = -k_\omega \text{sign}(e_{22_{ar}}). \end{aligned} \tag{2.19}$$

The second equation is applied when the robot is reaching the target avoiding the problem of short baseline. The condition to switch from the first to the second equation is given by thresholding the value of the epipole e_{23}.

Keeping the control strategy in the same epipolar context has the advantage of providing full control of the position and orientation during the whole task. In previous epipolar approaches, a depth correction stage is carried out by commuting to a feature-based correlation approach with null rotational velocity [99], or by commuting to homography-based control [84]. So, two different approaches are used to solve the visual servoing task in the referred works. In comparison to pure homography-based approaches [51], [86], which depend on the observed plane, our approach is able to correct longitudinal error without an explicit position estimation. Additionally, our control scheme solves the singularity problem by using bounded input velocities while a direct motion to the target is carried out. This problem has been tackled in [99] by avoiding to reach the singularity using a particular motion strategy. The strategy prevents the singularity occurrence, but has the drawback that the robot goes away of the target and then, it performs a backward motion in a second step without orientation control. In [87], one of the control inputs is not computed when crossing the singularity. These aspects show the benefits of the described control scheme over previous related approaches.

2.4 Stability Analysis

First, it is worth noting that the servoing task must be accomplished by carrying out the two described steps. In the following, the stability of the tracking control law is analyzed in each step separately. Notice that both steps are independent in the sense that they are applied sequentially. In this analysis, we consider that enough number of image features of the target scene are visible in the camera's field of view during the navigation and that the robot starts in a general location.

Proposition 2.1. *The control law that combines the decoupling-based control (2.12) by commuting to the bounded control (2.13) whenever* $|\phi_2 - \psi_2| < n\pi + T_h$, *where* T_h *is a suitable threshold and* $n \in \mathbb{Z}$, *achieves global stabilization of the system (2.8). Moreover, global stabilization is achieved even with uncertainty in parameters.*

Proof. Firstly, stabilization of system (2.8) by using controller (2.12) is proved by showing that the sliding surfaces (2.10) can be reached in a finite time (existence conditions of sliding modes). Let us consider the natural Lyapunov function for a sliding mode controller

$$V_{st_1} = V_1 + V_2, \quad V_1 = \frac{1}{2}s_c^2, \quad V_2 = \frac{1}{2}s_t^2, \tag{2.20}$$

which accomplishes $V_{st_1}(s_c = 0, s_t = 0) = 0$ and $V_{st_1} > 0$ for all $s_c \neq 0$, $s_t \neq 0$. We analyze each term of the time-derivative

$$\dot{V}_{st_1} = \dot{V}_1 + \dot{V}_2 = s_c\dot{s}_c + s_t\dot{s}_t \tag{2.21}$$

for the decoupling-based controller (2.12). Using (2.8) for the time-derivatives of the sliding surfaces and the estimated parameters α_{x_e} and d_{23_e} in the controller, we have

$$\dot{V}_1 = s_c\left(-\frac{\alpha_x}{\alpha_{x_e}}\left(\kappa_c \mathrm{sign}\left(s_c\right) + \lambda_c s_c\right) + A\right) = -\left(\frac{\alpha_x}{\alpha_{x_e}}\left(\kappa_c\left|s_c\right| + \lambda_c s_c^2\right) - s_c A\right),$$

$$\dot{V}_2 = s_t\left(-\frac{\alpha_x d_{23_e}}{\alpha_{x_e} d_{23}}\left(\kappa_t \mathrm{sign}(s_t) + \lambda_t \xi_t\right) + B\right) = -\left(\frac{\alpha_x d_{23_e}}{\alpha_{x_e} d_{23}}\left(\kappa_t\left|s_t\right| + \lambda_t s_t^2\right) - s_t B\right),$$

where $A = \frac{\alpha_x}{\alpha_{xe}}\left(\frac{d_{23e}}{d_{23}} - 1\right)\left(\dot{e}_{32}^d - \kappa_t \mathrm{sign}(s_t) - \lambda_t s_t\right)\frac{\cos^2(\psi_2)}{\cos^2(\phi_2 - \psi_2)} + \left(\frac{\alpha_x}{\alpha_{xe}} - 1\right)\dot{e}_{23}^d$ and $B = \left(\frac{\alpha_x d_{23_e}}{\alpha_{x_e} d_{23}} - 1\right)\dot{e}_{tx}^d$ are obtained from $\mathbf{M}\cdot\mathbf{M}_e^{-1}$. After some simplifications, we can see that

$$\dot{V}_1 \leq -\left(\frac{\alpha_x}{\alpha_{x_e}}\left(\kappa_c + \lambda_c\left|s_c\right|\right) - |A|\right)\left|s_c\right|, \quad \dot{V}_2 \leq -\left(\frac{\alpha_x d_{23_e}}{\alpha_{x_e} d_{23}}\left(\kappa_t + \lambda_t\left|s_t\right|\right) - |B|\right)\left|s_t\right|.$$

Thus, $\dot{V}_1$ and $\dot{V}_2$ are negative definite if and only if the following inequalities are guaranteed for all $s_c \neq 0$, $s_t \neq 0$:

$$\kappa_c + \lambda_c\left|s_c\right| > \frac{\alpha_{x_e}}{\alpha_x}|A|, \quad \kappa_t + \lambda_t\left|s_t\right| > \frac{\alpha_{x_e} d_{23}}{\alpha_x d_{23_e}}|B|. \tag{2.22}$$

Therefore, $\dot{V}_{st_1} < 0$ if and only if both inequalities (2.22) are fulfilled. On one hand, it is clear that for ideal conditions $d_{23_e} = d_{23}$, $\alpha_{x_e} = \alpha_x$, the right side of both inequalities is zero and therefore, any control gains $\kappa_c > 0$, $\kappa_t > 0$, $\lambda_c > 0$, $\lambda_t > 0$, accomplish the inequalities. On the other hand, when the estimated controller parameters are different to the real ones, the right side of inequalities become proportional to $\left|\dot{e}_{cx}^d\right|$, $\left|\dot{e}_{tx}^d\right|$. By using slow references and increasing slightly the gains, global convergence to the sliding surfaces can be achieved regardless of uncertainty in parameters.

Now, let us show the stabilization of system (2.8) by reaching the surfaces (2.10) using the controller (2.13). The same Lyapunov function (2.20) is used, and for each term of (2.21) we have

$$\dot{V}_1 = -k_\omega \frac{\alpha_x}{\cos^2(\phi_2 - \psi_2)} |s_c| - s_c \dot{e}_{cx}^d - s_c C,$$
$$\dot{V}_2 = -k_\upsilon \frac{\alpha_x |b(\phi_2, \psi_2)|}{d_{23} \cos^2(\psi_2)} |s_t| - s_t \dot{e}_{tx}^d,$$

where $C = k_\upsilon \frac{\alpha_x |b(\phi_2, \psi_2)|}{d_{23} \cos^2(\phi_2 - \psi_2)} \operatorname{sign}(s_t)$ and $b(\phi_2, \psi_2)$ is given in (2.14). So, we obtain

$$\dot{V}_1 \le -\left(k_\omega \frac{\alpha_x}{\cos^2(\phi_2 - \psi_2)} - \left|\dot{e}_{23}^d\right| - |C|\right) |s_c|,$$
$$\dot{V}_2 \le -\left(k_\upsilon \frac{\alpha_x |b(\phi_2, \psi_2)|}{d_{23} \cos^2(\psi_2)} - \left|\dot{e}_{32}^d\right|\right) |s_t|.$$

It can be verified that $\dot{V}_1$ and $\dot{V}_2$ are negative definite if and only if the following inequalities are assured for all $s_c \neq 0$, $s_t \neq 0$:

$$k_\omega > \frac{\cos^2(\phi_2 - \psi_2)}{\alpha_x} \left(|C| + \left|\dot{e}_{23}^d\right|\right), \quad k_\upsilon > \frac{d_{23} \cos^2(\psi_2)}{\alpha_x |b(\phi_2, \psi_2)|} \left|\dot{e}_{32}^d\right|. \tag{2.23}$$

It is worth noting that the denominator of the right hand side of the last inequality does not become null because of the real behavior of sliding mode control that drives the system inside a tight band around the sliding surface, without the possibility of maintaining the system on the surface, which could only happens in ideal conditions [138]. Moreover, as mention above the inequalities (2.23), the proof of asymptotic stability with finite convergence time excludes the occurrence of $b = 0$ given that $s_c \neq 0$. Therefore, $\dot{V}_{st_1} < 0$ if and only if both inequalities (2.23) are fulfilled. Once again, these inequalities are accomplished by using slow desired trajectories for the epipoles with appropriate gains. Note that these inequalities do not depend on the controller parameters α_{x_e}, d_{23_e} because the bounded controller does not need any information of system parameters and thus, its robustness is implicit.

The bounded controller (2.13) is able to locally stabilize the system (2.8) and its region of attraction grows as long as the control gains k_υ and k_ω are higher. Given that the control strategy commutes between two switching controllers according to a rule defined by the threshold T_h (so that each one acts inside of its region of attraction), the commutation between them does not affect the stability of the overall control system.

Once sliding surfaces are reached for any case of sliding mode control law, the system's behavior is independent of matched uncertainties and disturbances. Uncertainties in the system (2.8) fulfill the matching condition; they belong to the range space of the input vector, and as a result, robustness of the control loop is accomplished. □

Proposition 2.2. *The translational velocity υ_{dc} (2.18) achieves global stabilization of the system (2.16) even with uncertainty of parameters, while the rotational velocity (2.19) achieves lateral drift compensation assuming that Proposition 2.1 is accomplished.*

Proof. We prove the stabilization of system (2.16) by using controller (2.18) and showing that the sliding surfaces (2.17) can be reached in a finite time. Simultaneously, given the control action (2.19), the epipoles e_{23} and respectively $e_{22_{ar}}$ are maintained in zero with finite time convergence, keeping the alignment with the target. Let us define the following candidate Lyapunov function

$$V_{st_2} = \frac{1}{2}s^2 + \frac{1}{2}\left(e_{2\star}^2 + e_{\star 2}^2\right),$$

where $\star$ refers to the target $I_3(\mathbf{K}, \mathbf{C}_3)$ or the intermediate image $I_{2_{ar}}(\mathbf{K}, \mathbf{C}_{2_{ar}})$. The time-derivative of this function is

$$\dot{V}_{st_2} = s\dot{s} + e_{2\star}\dot{e}_{2\star} + e_{\star 2}\dot{e}_{\star 2}.$$

The dynamic $\dot{s}$ is obtained from (2.16) and $\dot{e}_{2\star}$, $\dot{e}_{\star 2}$ are given as

$$\begin{aligned} \dot{e}_{2\star} &= \frac{-\alpha_x \sin(\phi_2 - \psi_2)}{d_{2\star}\cos^2(\phi_2 - \psi_2)}\upsilon_{dc} + \frac{\alpha_x}{\cos^2(\phi_2 - \psi_2)}\omega_b, \\ \dot{e}_{\star 2} &= \frac{-\alpha_x \sin(\phi_2 - \psi_2)}{d_{2\star}\cos^2(\psi_2)}\upsilon_{dc}. \end{aligned} \tag{2.24}$$

By the assumption that Proposition 2.1 is accomplished, the robot starts the second step aligned with the target ($x_2 = 0$, $\phi_2 = 0$), which implies $\phi_2 - \psi_2 \approx 0$. Then, we use the small angle approximation $\sin(\phi_2 - \psi_2) \approx 0$, $\cos(\phi_2 - \psi_2) \approx 1$ to obtain

$$\dot{V}_{st_2} = -k_1 \frac{\alpha_x d_{12e}}{\alpha_{x_e} d_{12}} |s| + Ds - k_\omega \alpha_x |e_{2\star}|,$$

where $D = \left(1 - \frac{\alpha_x d_{12e}}{\alpha_{x_e} d_{12}}\right) k_0 \xi_{12}$. Therefore, $\dot{V}_{st_2}$ is negative definite if and only if the following inequalities are guaranteed for all $s \neq 0$ and $e_{2\star} \neq 0$:

$$k_1 > \frac{\alpha_{x_e} d_{12}}{\alpha_x d_{12e}} |D|, \; k_\omega > 0. \tag{2.25}$$

For ideal conditions, the right side of the first inequality is zero and any value $k_1 > 0$ is enough to reach the sliding surface in finite time. On the contrary, when controller parameters are different to the real ones, the gain k_1 should be increased. Once the sliding mode is reached, the stability of the reduced order system is guaranteed for $k_0 > 0$. Additionally, any disturbance caused by the small angle approximation accomplishes the matching condition and it can be rejected by the sliding mode control input. So, the system (2.24) is maintained around $e_{2\star} = 0$, $e_{\star 2} = 0$ and the alignment to the target ($\phi_2 = 0$, $x_2 = 0$) is ensured correcting any possible deviation. Finally, the joint action of υ_{dc} (2.18) and ω_b (2.19) steers the robot in straight motion toward the target in the second step. The stop condition $e_{12} - e_{13} = 0$ guarantees to reach the desired location $(x_2 = 0, y_2 = 0, \phi_2 = 0)$. □

Note that the parameters d_{23} and d_{12} are unknown, but according to conditions (2.22) and (2.25), they appear as a factor of the translational velocity that can be absorbed by the control gains. However, a good strategy to set the corresponding controller parameters d_{23_e} and d_{12_e} is to over-estimate them, ensuring that they are coherent with the scenario.

Although we are not dealing with a totally uncalibrated case, we have shown that robust global stabilization of the error function can be achieved by setting adequate control gains. Our approach has been developed specifically for mobile robots on the basis of a square control system unlike some uncalibrated approaches, for instance [93]. Indeed, our approach does not require any additional feedback or initialization information, in contrast with [140].

2.5 Experimental Evaluation

An extensive evaluation of the previously described visual control scheme is presented in the sequel of the chapter. Results show that the main objective of driving the robot to a desired pose ((0,0,0^o) in all the cases) is attained regardless of passing through the singularity that occurs in the first step for some initial poses, and moreover, the task is accomplished even when the robot starts exactly in a singular pose.

2.5.1 Simulation Results

In this section, we present some simulations of our visual control. Simulations have been performed in Matlab with a sampling time of 100 ms. We report realistic results using simulated conventional and central catadioptric cameras. As mentioned previously, the scheme is calibrated for central catadioptric cameras. However, for the case of conventional cameras, we present the good performance and robustness of the approach under a wide range of parametric uncertainty. Finally, a simulation with a complete dynamic model of the robot is presented.

Along the first two subsequent sections we keep fixed all the parameters of the control law. We set an estimated focal length (f_e) of 6 mm, an initial distance between the current and target positions (d_{23_e}) of 10 m, the same for d_{12_e}. The duration of the first step (alignment with the target) is fixed to 80% of the task execution time and the remaining time is left for longitudinal error correction. The threshold to commute to the bounded control law (T_h) is fixed to 0.03 rad. Related to the control gains, they are set to $\lambda_c = 2$, $\lambda_t = 1$, $\kappa_c = 2$, $\kappa_t = 2$, $k_\upsilon = 0.1$, $k_\omega = 0.06$, $k_0 = 1$ and $k_1 = 2$.

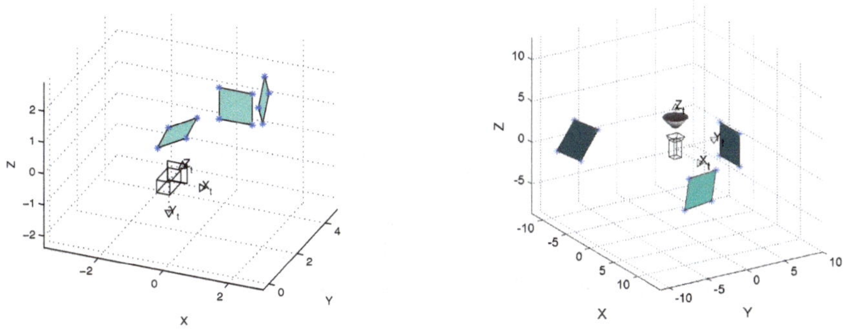

(a) (b)

Fig. 2.4 3D virtual scenes used to generate synthetic images. (a) Conventional camera. (b) Central catadioptric camera. Draw of cameras with EGT ([101]).

2.5.1.1 Evaluation Using Conventional Cameras

In this section, we use virtual perspective images of size 640×480 pixels, which are generated from the 3D scene shown in Fig. 2.4(a). The camera is fixed to the robot looking forward and the task execution time is set to 20 s. Fig. 2.5(a) presents the resultant paths and Fig. 2.5(b) the epipoles evolution for the initial poses (4,-8,40^o), (3,-11,15.26^o), (-5,-13,0^o) and (-7,-18,-10°).

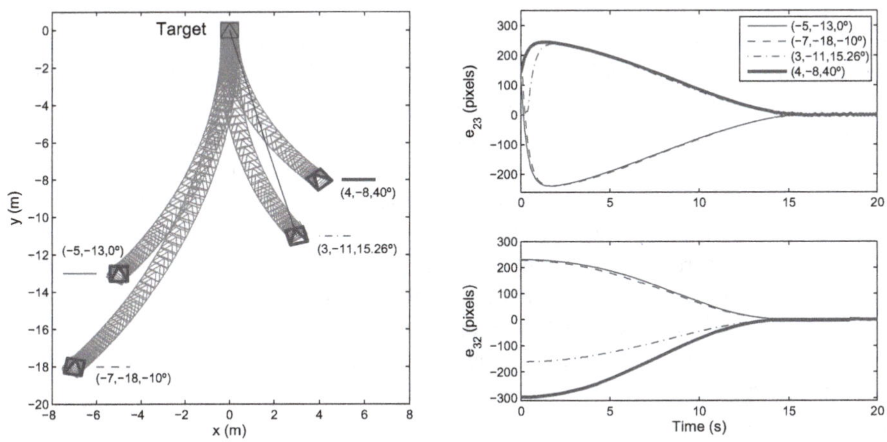

(a) (b)

Fig. 2.5 Simulation results with conventional cameras for different initial locations. (a) Paths on the $x-y$ plane. (b) Current and target epipoles.

In the two last cases the plots of the epipoles are almost superimposed, the robot starts with $\text{sign}(e_{23}) = \text{sign}(e_{32})$ and the epipoles are taken to the desired trajectories. In both cases e_{23} changes its sign during the first seconds, which causes a rotation of the robot, and then, it begins a direct motion toward the target. The initial pose (3,-11,15.26^o) corresponds to a special case where the state trajectory just starts on the singularity $e_{23} = 0$. The line from the robot initial position to the target shows that the camera axis is aligned with the baseline for this pose. When the robot starts just on the singularity, we assign a suitable amplitude to the desired trajectory for the current epipole. Given that $|\phi_2 - \psi_2|$ is less than the threshold, the bounded controller takes the system out of the singularity and then, the epipoles evolve as shown in Fig. 2.5(b).

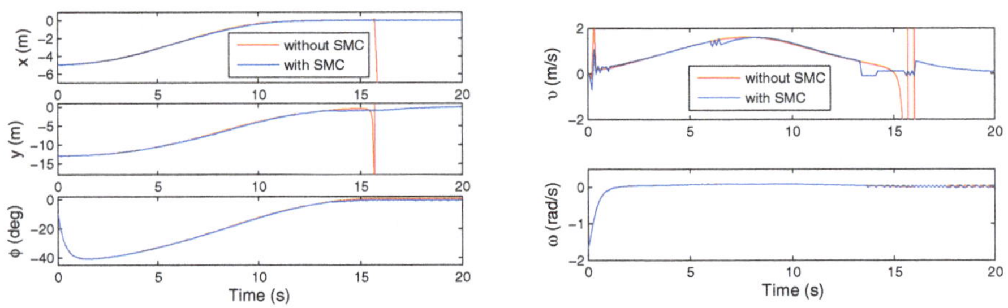

(a) (b)

Fig. 2.6 Simulation results with conventional cameras for a case where the singularity is crossed at the beginning ((-5,-13,0^o) of Fig. 2.5). (a) State of the robot. (b) Control inputs.

Note that the evolution of the epipoles crosses the singularity $e_{23} = 0$ for the initial cases (-5,-13,0^o) and (-7,-18,-10°). The behavior of the state of the robot is presented in Fig. 2.6(a) for the first case. We have included the behavior without using sliding mode control to show the explosion of the state at the singularity. The corresponding computed velocities are presented in Fig. 2.6(b), where can be seen that the sliding mode control provides bounded velocities. The control inputs are maintained bounded even when the epipoles are close to zero around 16 s, which ensures entire correction of orientation and lateral position. We can see an exponential decay for the translational velocity after 16 s, which corrects any remaining longitudinal error. The good behavior of the approach can be seen also in the image space. Fig. 2.7 shows the motion of the point features in the virtual images. We can notice that the images at the end of the motion represented by the times signs are practically the same as the target images represented by the red circles.

2.5.1.2 Evaluation Using Omnidirectional Cameras

In this section, we present the behavior of our control scheme using virtual omnidirectional images of size 640×480 pixels, which are generated from the 3D scene shown in Fig. 2.4(b). In this case, the camera calibration parameters are needed in

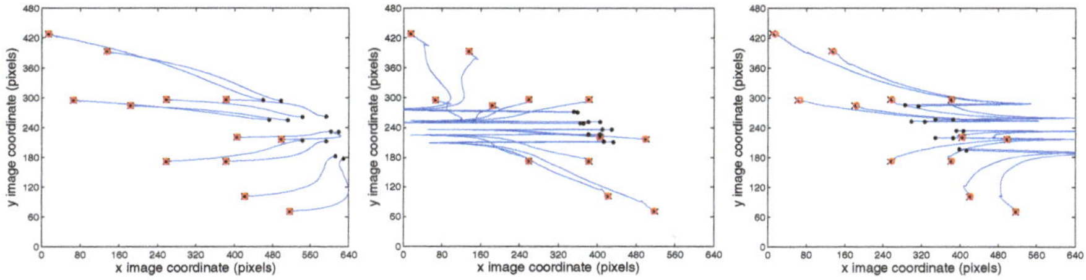

(a) (b) (c)

Fig. 2.7 Motion of the image points using conventional cameras for initial locations: (a) (4,-8,40^o), (b) (-7,-18,-10°), (c) (3,-11,15.26^o). The black points represent the initial image, the red circles represent the target image and the times signs represent the image at the end of the motion.

order to obtain the points on the unitary sphere, from which, the epipolar geometry can be estimated similarly as for conventional cameras. Results using different types of central cameras are included. Moreover, we also report results using fisheye cameras, which in spite of being noncentral imaging systems, their projection process may be well approximated by the generic central camera model with the adequate parameters. The task execution time has been set to 60 s.

Similar results to the previous section are obtained by using omnidirectional vision, but with the additional advantage of ensuring to keep the features in the field of view during the servoing. Fig. 2.8(a) shows the resultant robot motion and Fig. 2.8(b) depicts the epipoles evolution for the initial poses (-5,-9,-50°), (-4,-14,0^o), (8,-16,10^o) and (2.5,-12,11.77^o). The first case corresponds to a position from where the robot can exert a direct navigation to the target and has been tested using a hypercatadioptric camera. In the second and third cases, the robot rotates initially and then it begins a direct motion toward the target after crossing the singularity. These cases are tested using paracatadiotric and fisheye cameras respectively. The last case, tested using a hypercatadioptric imaging system, shows that the robot is effectively driven to the target from a singular initial pose, where $e_{23} = 0$.

Fig. 2.9(a) presents the behavior of the state of the robot for the case (2.5,-12,11.77^o), i.e., the singular initial pose. We can see in Fig. 2.9(b) that both of the input velocities are maintained bounded at the beginning and along the navigation. Given the time execution time of these simulations, the alignment with the target is reached in 48 s. After that, we can see the exponential behavior of the translational velocity, which corrects the remaining longitudinal error while orientation is also preserved through the bounded switching rotational velocity. The good behavior of the approach in the image space can be seen in Fig. 2.10, where the motion of the point features in the different types of virtual images is shown.

2.5.1.3 Robustness under Parametric Uncertainty

In this section, we report some tests in order to show the robustness of the control law under uncertainty in parameters for conventional cameras, where calibration

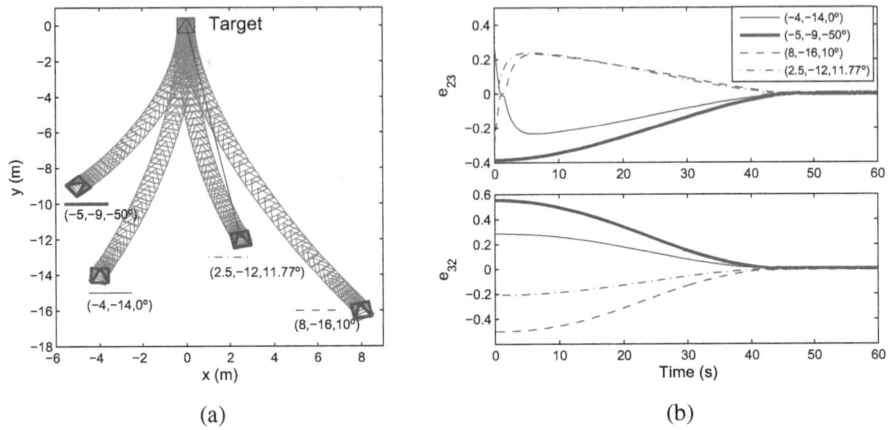

Fig. 2.8 Simulation results with omnidirectional cameras for different initial locations. (a) Paths on the $x-y$ plane. (b) Current and target epipoles.

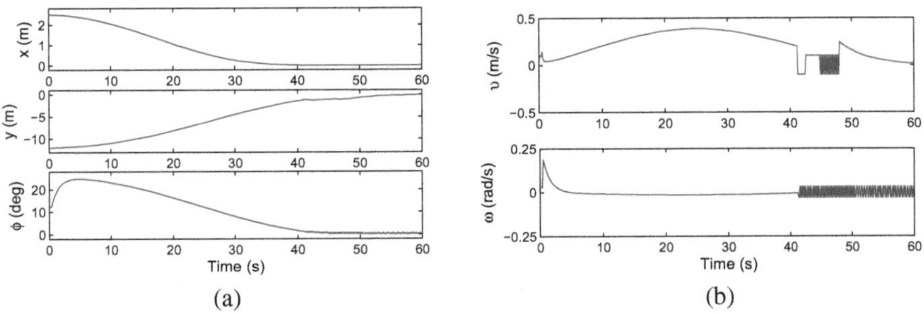

Fig. 2.9 Simulation results with omnidirectional cameras for a case where the robot starts in a singular configuration ((2.5,-12,11.77^o) of Fig. 2.8). (a) State of the robot. (b) Control inputs.

can be avoided in contrast with the case of omnidirectional cameras. Up to now, all the parameters for the control law are maintained as described in section 2.5.1. The first aspect to notice is that for the four paths shown previously, the distance (d_{23} in meters) between cameras is different for each initial pose and even so, the final target is reached with good precision for all the cases (see Table 2.1).

Regarding to camera parameters uncertainty, we analyze the effect of changing the focal length (f) in the computation of epipoles while keeping f_e constant in the controller. The initial position is (2,-7,30^o) for all the cases, however, the obtained behavior is recurrent for any initial pose. Fig. 2.11 presents the final pose and mean squared tracking error for a wide range of focal length values. We can see that regardless of the difference of the focal length used in the controller with respect

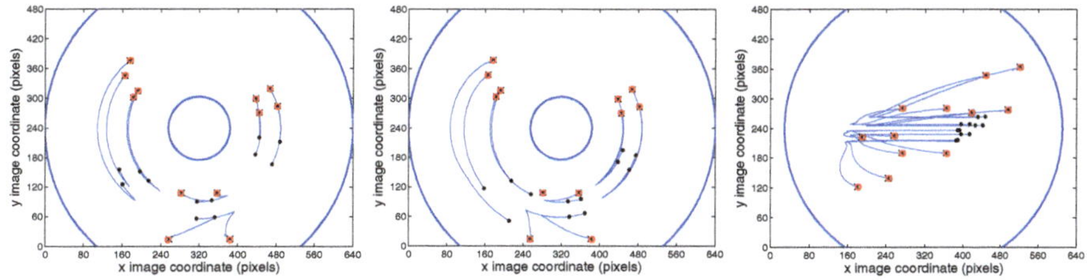

(a) (b) (c)

Fig. 2.10 Motion of the points in the image plane for different omnidirectional images. (a) (2.5, -12,11.77^o) Hypercatadioptric. (b) (-4,-14,0^o) Paracatadioptric. (c) (8,-16,10^o) Fisheye. The black points represent the initial image, the red circles represent the target image and the times signs represent the image at the end of the motion.

Table 2.1 Robustness under different initial distance between cameras ($d_{23} = d_{12} = d$)

	$d = 8.9$ (4,-8)m	$d = 11.4$ (3,-11)m	$d = 13.9$ (-5,-13)m	$d = 19.3$ (-7,-18)m
x_{final} (cm)	0	0	0	0
y_{final} (cm)	0.53	0.16	−2.84	−4.10
ϕ_{final} (deg)	−0.08	0.04	0.01	−0.02

to the real one, the robot always reaches the target with good precision and the tracking error is maintained in a low value. The last plot in Fig. 2.11 shows the final pose for different values of the x-coordinate of the principal point. In all the trials the target is reached closely.

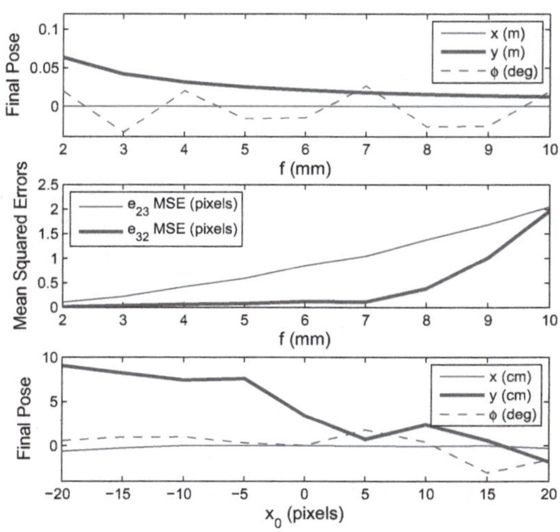

Fig. 2.11 Simulations with different values of focal length (f) and principal point (x_0) showing robustness against parametric uncertainty

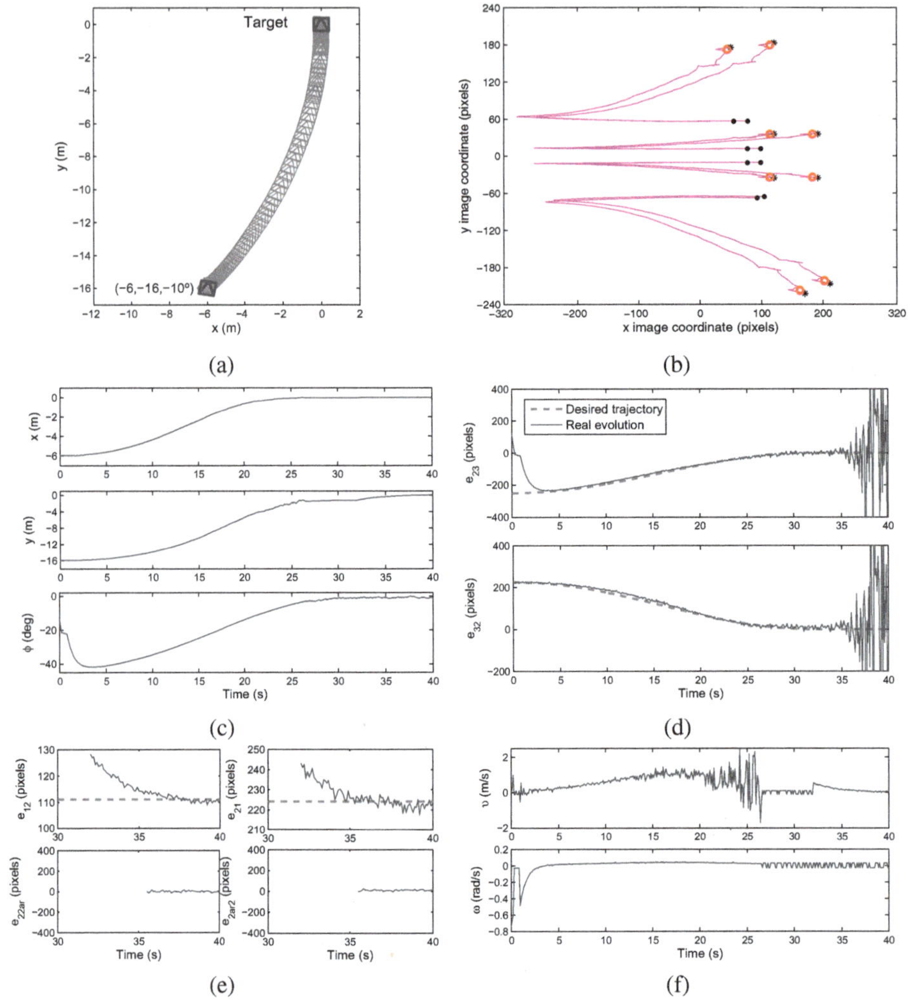

Fig. 2.12 Simulation results: robustness under image noise. (a) Robot trajectory on the $x-y$ plane. (b) Motion of the points in the image. The black points represent the initial image, the red circles represent the target image and the black asterisks represent the image at the end of the motion. (c) State variables of the robot during the motion. (d) Epipoles e_{23} and e_{32}. (e) Epipoles e_{12}, e_{21}, $e_{22_{ar}}$ and $e_{2_{ar}2}$. (f) Computed velocities.

Fig. 2.12(a) shows the performance of the approach under image noise for the initial pose (-6,-16,-10°). The simulation time is set to 40 s and the noise added to the image points has a standard deviation of 0.5 pixels. It is clear the presence of this noise in the motion of the image points in Fig. 2.12(b). In Fig. 2.12(c) we can see the exponential behavior of the depth after $T = 32$ s, which reaches zero by using feedback from e_{12}. We can notice in Fig. 2.12(d) that the epipoles e_{23} and

e_{32} become unstable before the end. However, after 32 s the controller uses e_{12} to compute the translational velocity by regulating e_{12} to a constant value as shown in Fig. 2.12(e). We can see that e_{21} is more sensitive because also depends on the rotational velocity, but it is not used in the controller.

The corresponding input velocities obtained by the control algorithm are shown in Fig. 2.12(f). The epipoles e_{23} and e_{32} are used to compute the decoupled velocities (2.12) until 26 s and the bounded velocities (2.13) until 32 s. The same rotational velocity computed from e_{23} is maintained from 32 s to 35.5 s. As e_{23} turns out unstable, from 35.5 s to 40 s, $e_{22_{ar}}$ is used to compute the rotational velocity according to (2.19). The translational velocity between 32 s and 40 s is computed from e_{12} given by (2.18). Notice that none of both velocities are subject to the problem of short baseline at the end of the motion, since they are computed from stable measurements (e_{12} and $e_{22_{ar}}$, respectively).

2.5.1.4 Complete Dynamic Robot Model

The next simulation has been performed with a complete dynamic model of the vehicle to show the performance of the controller in this realistic situation. For this purpose we use Webots^{TM} (http://www.cyberbotics.com) [109], a commercial mobile robot simulation software developed by Cyberbotics Ltd. The physics simulation of Webots^{TM} relies on ODE (Open Dynamics Engine, http://www.ode.org)

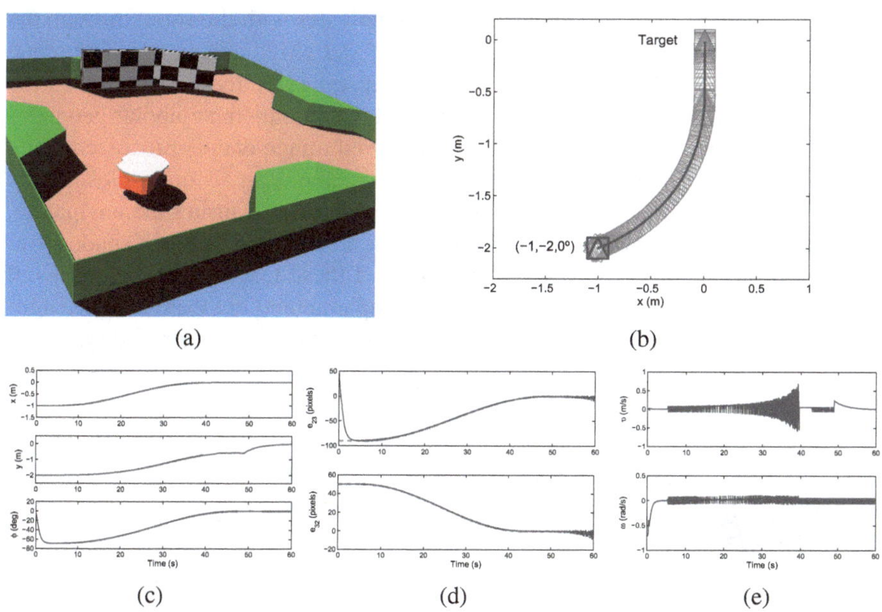

Fig. 2.13 Simulation results using a complete dynamic model for a robot Pioneer. (a) Simulation setup. (b) Path on the $x-y$ plane. (c) State of the robot. (d) Evolution of the epipoles. (e) Computed velocities.

to perform accurate physics simulation. We can define, for each component of the robot, parameters like mass distribution, static and kinematic friction coefficients, bounciness, etc.

In this experiment, we use the model of a robot Pioneer P3-DX (Fig. 2.13(a)) with the dynamic parameters, and the closed loop frequency is set to 10 Hz. In Fig. 2.13(b)-(d) it is noticeably that the chattering effect yielded by the control inputs in Fig. 2.13(e) is practically negligible. Chattering is an undesirable phenomenon presented in sliding mode control systems that generates an oscillation within a neighborhood of the switching surface such that $\mathbf{s} = \mathbf{0}$ is not satisfied as expected ideally. Although there exist methods for chattering suppression [76], if the frequency of the switching is high enough compared with the dynamic response of the system, then the chattering is not significative. In our results, the discontinuous switching control inputs have a relatively high frequency and because of the low-pass filtering effect of the robotic mechanical system, the state of the robot behaves smoothly.

2.5.2 *Real-World Experiments*

The complete control law of section 2.3 has been tested in real conditions using a robot Pioneer P3-DX from Adept Mobile Robots. The robot is equipped with a USB camera mounted on top (Logitech QuickCam Communicate STX) as shown in Fig. 1.1(a). The images are acquired at size 640×480 pixels. The camera is connected to a laptop onboard the robot (Intel® Core^{TM} 2 Duo CPU at 2.50 GHz) with operating system Debian Linux. This computer communicates with the robot through the serial port using the ARIA library available from ActivMedia Robotics. The scene observed is set up with two planes consisting on square patterns, from which the corners of the squares are extracted and matched to estimate the epipolar geometry relating the current and target images. The acquired image data is processed using the OpenCV library. This framework allows us to achieve an adequate closed loop frequency (limited to 100 ms due to hardware constraints). During the navigation, the system performs the tracking of the image points using a Lucas-Kanade pyramidal algorithm [92]. The corresponding points of these features are the entries of the 8-point epipolar computation algorithm as implemented in OpenCV. This algorithm automatically applies data normalization, solves the overdetermined system of

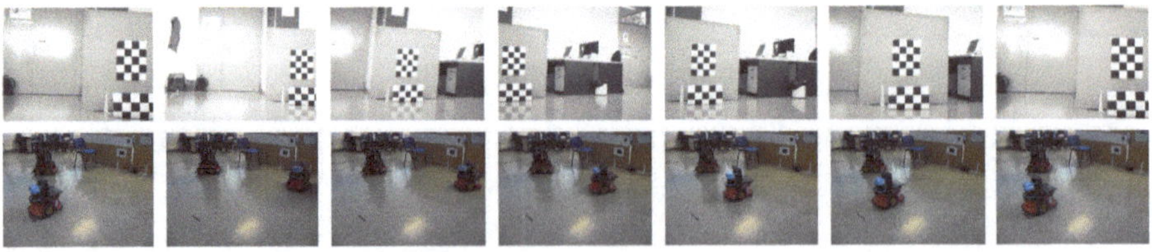

Fig. 2.14 Sequence of some images taken from the robot camera (1st row) and from an external camera (2nd row) during the real experiment. The first is the target image, the second is the initial and the last is the image at the end of the motion. The robot behind is not involved in the experiment.

equations built from the epipolar constraint and returns the adequate denormalized rank-2 fundamental matrix. Then the epipoles are obtained using the singular value decomposition. The control law parameters have been set to $d_{23_e} = d_{12_e} = 5$ m and $f_e = 9$ mm, and the image center as the principal point without performing specific calibration. Fig. 2.14 shows sequences of some images taken by the robot camera and an external video camera respectively for one of the runs of the experiment.

Fig. 2.15(a) presents the resultant path, given by the robot odometry, from the initial location (-0.3 m,-1.3 m, 0°) for one of the experimental runs. Fig. 2.15(b) shows the evolution of the state during the 30 s in which the positioning task is carried out. The final position error is less than 5 cm and the orientation error is practically negligible. The time τ for the execution of the first step, alignment with the target, is set to 21 s. We can see in Fig. 2.15(c) how the bounded sliding mode control law is applied around 16 s due to the occurrence of the singularity. It avoids the unbounded growing of the translational velocity at the same time that longitudinal error correction continues. After 21 s the feedback for this correction is provided

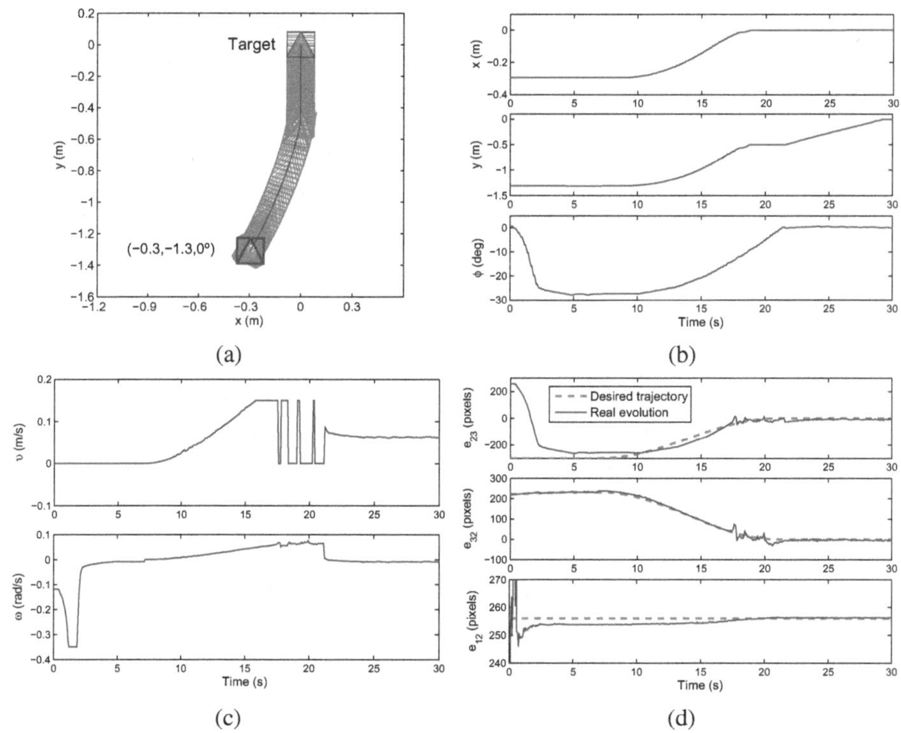

Fig. 2.15 Real experiment with target location (0,0,0^o). (a) Robot motion on the $x-y$ plane. (b) State of the robot. (c) Computed velocities. (d) Evolution of the epipoles involved in the control law. The data presented in (a)-(b) corresponds to the robot odometry. As can be seen in the plot of the linear velocity at the end, the robot moves forward until the termination condition explained after (2.18) is met and the robot stops.

from the error of e_{12}. The behavior of those epipoles involved in the control law is shown in Fig. 2.15(d). Notice that each one of them reaches its desired final values. Although the tracking error for the current epipole (e_{23}) is not as good as in simulations, the behavior is as expected. This epipole starts at a positive value and goes to negative during the initial rotation to reach finally the reference. The fluctuations in e_{23} and e_{32} around 20 s correspond to the switching behavior of the bounded translational input. However, note that this is not reflected on the state of the robot. The termination condition of the task is given when the difference $e_{12} - e_{13}$ is less than a threshold.

The non-ideal behavior of the tracking for e_{23} is due to the hardware constraints, given that the closed loop frequency is limited in the robots at our disposal. Nevertheless, simulations and real-world experiments show that a closed loop frequency around 10 Hz is enough to obtain system's behavior with small chattering effect. The experimental evaluation shows the validity of our proposal and its satisfactory performance with the hardware used. As long as the sampling period could be reduced the results could be better.

As can be noticed in the experimental evaluation of this section, we have focused on the problem of a direct robot motion toward the target. We do not consider explicitly maneuvers to be carried out. However, the pose regulation from some particular initial locations can be addressed by maneuvering, for instance, when the three cameras are aligned or there is only a lateral error. In those cases, the commuted control law of the first step is able to take the robot to a general configuration by defining adequate references for the epipoles. In this sense, the presented control law complies with the Brockett's theorem. The control inputs are time-varying and computed in two steps, in such a way that some maneuvers can be carried our if required.

2.6 Closure

In this chapter, a robust control law based on sliding mode control has been presented in order to perform image-based visual servoing of mobile robots. The approach is valid for differential-drive wheeled robots moving on a plane and carrying a conventional or an omnidirectional camera onboard. The generic control law has been designed on the basis of kinematic control by exploiting the pairwise epipolar geometries of three views. The interest of the ideas presented in this chapter turns out in a novel control law that performs orientation, lateral error and depth correction without needing to change to any approach other than epipolar-based control. The control scheme deals with singularities induced by the epipolar geometry maintaining always bounded inputs, which allows the robot to carry out a direct motion toward the target. Additionally, it is a robust scheme that does not need a precise camera calibration in the case of conventional cameras, although it is calibrated for omnidirectional vision. On one hand, the approach can be used provided that there are enough point features to estimate the epipolar geometry between the views. On the other hand, the sliding mode control requires an adequate closed loop frequency, which can be achieved in typical experimental hardware as shown in our real-world experiments.

Chapter 3
A Robust Control Scheme Based on the Trifocal Tensor

Abstract. In this chapter, we rely on the natural geometric constraint for three views, the trifocal tensor. We present a novel image-based visual servoing scheme that also solves the pose regulation problem, in this case by exploiting the properties of omnidirectional images to preserve bearing information. This is achieved by using the additional information of a third image in the geometric model through a simplified trifocal tensor, which can be computed directly from image features avoiding the need of a complete camera calibration for any type of central camera. The main idea of the chapter is that the elements of the tensor are introduced directly in the control law and neither any a prior knowledge of the scene nor any auxiliary image are required. Additionally, a sliding mode control law in a square system ensures stability and robustness for the closed loop. The good performance of the control system is proven via simulations and real-world experiments with a hypercatadioptric imaging system.

3.1 Introduction

Typically, the visual information to carry out visual servoing is extracted from two images: the target and the image acquired at the current location. We take advantage of more information by using three views and a particular geometric constraint that relates them. Besides the target and current views, it is always possible to save the initial image in order to exploit the trifocal tensor computed from that triplet of images. The trifocal tensor describes all the geometric relationships between three views and is independent of the observed scene [60].

The first work that proposes a robotic application of a trilinear constraint is [49], in which a simplified tensor is exploited, the so-called 1D trifocal tensor. In that work, conventional perspective cameras are converted to 1D virtual cameras through a transformation of bearing measurements for localization. In the context of computer vision, the same idea is introduced to wide-angle cameras as a tool for calibrating the radial distortion in [133]. The same authors present a general hybrid trifocal constraint by representing conventional and omnidirectional cameras as radial 1D

H.M. Becerra and C. Sagüés, *Visual Control of Wheeled Mobile Robots*,
Springer Tracts in Advanced Robotics 103,
DOI: 10.1007/978-3-319-05783-5_3, © Springer International Publishing Switzerland 2014

cameras in [134]. They assert that the radial 1D camera model is sufficiently general to represent the great majority of omnidirectional cameras under the assumption of knowing the center of radial distortion. The effectiveness of applying the 1D trifocal tensor to recover location information has been also proved in [59]. It uses the trifocal tensor with both conventional and omnidirectional cameras for scene reconstruction, and proposes this approach for initialization of bearing-only SLAM algorithms. A recent work presents a visual control for mobile robots based on the elements of a 2D trifocal tensor constrained to planar motion [85]. This approach shows good results reaching the target location, but it uses a non-exact system inversion that suffers of potential stability problems. Moreover, the benefits of using more than three views and higher order tensors have been explored for visual odometry [38].

In this chapter, we present an image-based visual servoing scheme for mobile robots that exploits the 1D trifocal tensor to define an adequate error function. The control law uses direct feedback of the elements of the 1D trifocal tensor. This idea has been introduced in [18], where a visual control based on the 1D trifocal tensor obtained from metric information is proposed for conventional cameras. However, because of the constrained field of view of conventional cameras, it is better to take advantage of the omnidirectional vision. Such extension has been developed in [14]. The approach is suitable for all central catadioptric cameras and even for fisheye cameras, since all of these imaging systems present high radial distortion but they preserve the bearing information, which is the only required data in our approach.

As detailed along the chapter, the approach does not require any a priori knowledge of the scene and does not need any auxiliary image. The control scheme ensures total correction of the robot pose even for initial locations where epipolar geometry or homography based approaches fail, for instance, avoiding the problem of short baseline. In comparison with classical image-based visual servoing schemes, the described scheme allows to prove stability of the closed loop on the basis of a square control system. Additionally, from a control theory point of view, we improve the robustness properties of the system by using sliding mode control. We show the robustness of the control law under image noise and the general performance is also analyzed through real-world experiments with images of a hypercatadioptric system.

The chapter is organized as follows. Section 3.2 details the 1D trifocal tensor and analyzes the possibilities to define an adequate error function from this constraint. Section 3.3 details the design procedure of the control law. Section 3.4 presents the stability analysis. Section 3.5 shows the performance of the control system via simulations with synthetic images, experimental analysis with real images and real-world experiments in closed loop. Finally, Section 3.6 closes the chapter with some conclusions.

3.2 Defining a Control Framework with the 1D Trifocal Tensor

The 1D trifocal tensor is a simplified tensor that relates three views in the frame of planar motion, which is the typical situation in the context of mobile robots. This tensor provides the advantage of being estimated from bearing visual measurements avoiding the need of complete camera calibration. In general, the point features have to be converted to their projective formulation in a 1D virtual retina in order to estimate the 1D trifocal tensor. The computation of this geometric constraint is basically the same for conventional cameras and for central catadioptric systems assuming that all of them approximately obey the generic central camera model of section 1.3.2.

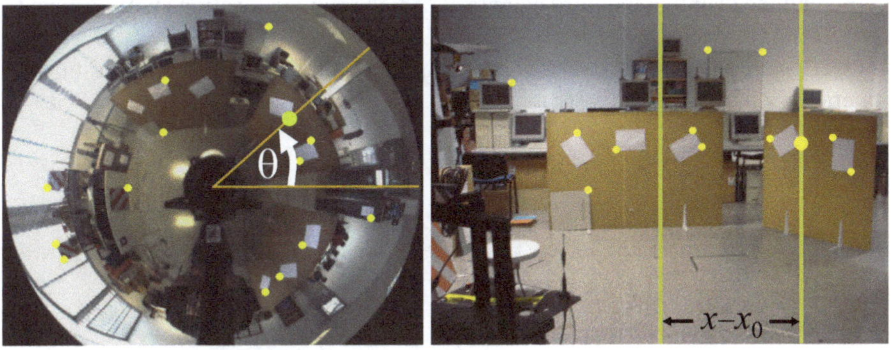

Fig. 3.1 Extracted measurements from central cameras to estimate the 1D trifocal tensor. (a) Hypercatadioptric image. (b) Perspective image.

For catadioptric imaging systems looking upward, this tensor particularly adapts to the property of these omnidirectional images to preserve bearing information in spite of the high radial distortion induced by lenses and mirrors. Fig. 3.1(a) shows the bearing angle of an observed feature in a hypercatadioptric system. The angle is measured with respect to a frame centered in the principal point of the image. For conventional cameras looking forward, the 1D projective formulation can be obtained as shown in Fig. 3.1(b) using the normalized x-coordinate of the point features with respect to x_0, i.e., $\mathbf{p} = \begin{bmatrix} x_n & 1 \end{bmatrix}^T$. For omnidirectional cameras, a bearing measurement θ can be converted to its 1D projection as $\mathbf{p} = \begin{bmatrix} \sin\theta & \cos\theta \end{bmatrix}^T$. By relating this representation for three different views of a feature that is expressed in a 2D projective space, it results in the simplified trifocal constraint

$$\sum_{i=1}^{2}\sum_{j=1}^{2}\sum_{k=1}^{2} T_{ijk}\mathrm{u}_i\mathrm{v}_j\mathrm{w}_k = 0, \tag{3.1}$$

where $\mathbf{u} = [\mathrm{u}_1,\mathrm{u}_2]^T$, $\mathbf{v} = [\mathrm{v}_1,\mathrm{v}_2]^T$ and $\mathbf{w} = [\mathrm{w}_1,\mathrm{w}_2]^T$ are the image coordinates of a feature projected in the 1D virtual retina of the first, second and third camera respectively, and T_{ijk} are the eight elements of the 1D trifocal tensor.

The described representation of bearing measurements is sufficiently general to model from pin-hole cameras to omnidirectional ones, as shown in [134]. Moreover, it allows computing a mixed trifocal constraint for heterogeneous cameras. In our case, the three images are captured by the same omnidirectional camera. In order to compute the eight elements of the 1D trifocal tensor we have to solve the linear system of equations obtained from seven stacked trifocal constraints (3.1). Thus, seven triples of matched features (eventually five for the calibrated case) are required to estimate the 1D trifocal tensor linearly.

Let us define a global (world) reference frame in the plane as depicted in Fig. 3.2(a) with the origin in the third camera. Then, the camera locations with respect to that global reference are $\mathbf{C}_1 = (x_1, y_1, \phi_1)$, $\mathbf{C}_2 = (x_2, y_2, \phi_2)$ and $\mathbf{C}_3 = (0,0,0)$. The relative locations between cameras are defined by a local reference frame in each camera as is shown in Fig. 3.2(b).

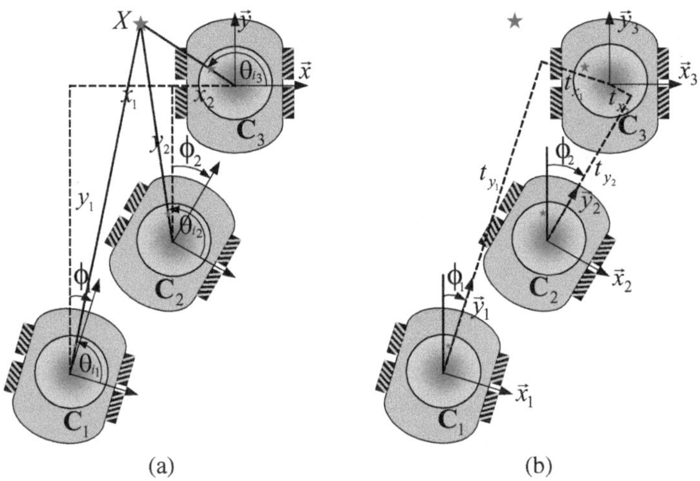

Fig. 3.2 Complete geometry between three camera-robot locations in the plane. (a) Absolute locations and bearing measurements extracted from omnidirectional images. (b) Relative locations.

The geometry of the three described views is encoded in the tensor elements as follows:

$$\mathbf{T}^m_{ijk} = \begin{bmatrix} T^m_{111} \\ T^m_{112} \\ T^m_{121} \\ T^m_{122} \\ T^m_{211} \\ T^m_{212} \\ T^m_{221} \\ T^m_{222} \end{bmatrix} = \begin{bmatrix} t_{y_1}\sin\phi_2 - t_{y_2}\sin\phi_1 \\ -t_{y_1}\cos\phi_2 + t_{y_2}\cos\phi_1 \\ t_{y_1}\cos\phi_2 + t_{x_2}\sin\phi_1 \\ t_{y_1}\sin\phi_2 - t_{x_2}\cos\phi_1 \\ -t_{x_1}\sin\phi_2 - t_{y_2}\cos\phi_1 \\ t_{x_1}\cos\phi_2 - t_{y_2}\sin\phi_1 \\ -t_{x_1}\cos\phi_2 + t_{x_2}\cos\phi_1 \\ -t_{x_1}\sin\phi_2 + t_{x_2}\sin\phi_1 \end{bmatrix}, \tag{3.2}$$

where $t_{x_i} = -x_i\cos\phi_i - y_i\sin\phi_i$, $t_{y_i} = x_i\sin\phi_i - y_i\cos\phi_i$ for $i = 1,2$ and the superscript m states that they are the tensor elements given by metric information. The complete deduction of the trifocal constraint (3.1) and the expressions (3.2) can be verified in [59]. There exist two additional constraints that are accomplished when the radial trifocal tensor is computed from a calibrated camera: $-T_{111} + T_{122} + T_{212} + T_{221} = 0$, and $T_{112} + T_{121} + T_{211} + T_{222} = 0$. These calibration constraints allow us to compute the 1D trifocal tensor from only five triplets of point correspondences, which improves the tensor estimation [59]. It is worth noting that these additional constraints can be always used for omnidirectional images because the bearing measurements are independent on focal length in that case. Therefore, to estimate de 1D trifocal tensor, we only require the center of projection for omnidirectional images or the principal point for conventional cameras.

In order to fix a common scale during the navigation, each estimated element of the tensor must be normalized dividing them by a non-null element ($\mathbf{T}_{ijk} = \mathbf{T}^e_{ijk}/T_N$). We can see from (3.2) that T_{121} tends to t_{y_1} as the robot reaches the target. If the initial robot location $\mathbf{C}_1$ is different to $\mathbf{C}_3$, we have $T_{121} \neq 0$. Additionally, this tensor element changes slightly as the robot moves. This fact is determined by the form of the derivative of T_{121}, which directly depends on the products $\omega\sin\phi_1$ and $\omega\sin\phi_2$, corresponding to small values in our framework. This is also supported by simulations and real experiments. Thus, in the sequel we assume that T_{121} is constant, and therefore, used as normalizing factor.

To design a controller for solving the pose regulation problem using only the tensor elements, we have to consider the corresponding final tensor values as control objective, analyze the dynamic behavior of the tensor elements and select an adequate set of them as outputs to be controlled. This analysis is carried out from the tensor elements as defined above.

3.2.1 Values of the 1D Trifocal Tensor in Particular Locations

Let us define the initial location of the robot to be (x_1, y_1, ϕ_1), the target location $(x_3, y_3, \phi_3) = (0,0,0)$ and (x_2, y_2, ϕ_2) the current location, which varies as the robot moves. It is worth emphasizing that $\mathbf{C}_1$ could be the moving camera and similar overall results may be obtained. Initially, when the second camera is in the starting location then $\mathbf{C}_2 = \mathbf{C}_1$, i.e., $(x_2, y_2, \phi_2) = (x_1, y_1, \phi_1)$, the relative location between these cameras is $t_{x_2} = t_{x_1}$, $t_{y_2} = t_{y_1}$ and the values of the tensor elements produce the relationships

$$T_{111} = 0,\ T_{112} = 0,\ T_{221} = 0,\ T_{222} = 0,\ T_{121} + T_{211} = 0,\ T_{122} + T_{212} = 0. \quad (3.3)$$

When the robot is in the goal $\mathbf{C}_2 = \mathbf{C}_3$, i.e., $(x_2, y_2, \phi_2) = (0,0,0)$, the relative location between these cameras is $t_{x_2} = 0$, $t_{y_2} = 0$, and it yields the relationships

$$T_{111} = 0,\ T_{122} = 0,\ T_{211} = 0,\ T_{222} = 0,\ T_{112} + T_{121} = 0,\ T_{212} + T_{221} = 0. \quad (3.4)$$

3.2.2 Dynamic Behavior of the Elements of the 1D Trifocal Tensor

In order to carry out the control from the tensor elements, we have to obtain the dynamic system that relates the change in the tensor elements exerted by a change in the velocities of the robot. This dynamic system involves the robot model (1.2) with $\ell = 0$ and is obtained by finding the time-derivatives of the tensor elements (3.2). We show two examples of the procedure to obtain these time-derivatives. The non-normalized tensor is denoted by T^m_{ijk}. From (3.2) and using the rates of change of the state variables (1.2) for T_{111} we have

$$\begin{aligned}
T^m_{111} &= t_{y_1} \sin\phi_2 - (x_2 \sin\phi_2 - y_2 \cos\phi_2) \sin\phi_1, \\
\dot{T}^m_{111} &= t_{y_1} \dot{\phi}_2 \cos\phi_2 - \left(\dot{x}_2 \sin\phi_2 + x_2 \dot{\phi}_2 \cos\phi_2 - \dot{y}_2 \cos\phi_2 + y_2 \dot{\phi}_2 \sin\phi_2\right) \sin\phi_1 \\
&= t_{y_1} \omega \cos\phi_2 - (-\upsilon \sin\phi_2 \sin\phi_2 + x_2 \omega \cos\phi_2 - \upsilon \cos\phi_2 \cos\phi_2 + y_2 \omega \sin\phi_2) \sin\phi_1 \\
&= \upsilon \sin\phi_1 + \omega \left(t_{y_1} \cos\phi_2 + t_{x_2} \sin\phi_1\right) = \upsilon \sin\phi_1 + T^m_{121} \omega.
\end{aligned}$$

By applying (1.10) in both sides of the equation, it results in the normalized time-derivative of T_{111}

$$\dot{T}_{111} = \frac{\sin\phi_1}{T^m_N} \upsilon + T_{121} \omega.$$

The same procedure is carried out for each element. Thus, for T_{121}

$$\begin{aligned}
T^m_{121} &= t_{y_1} \cos\phi_2 + (-x_2 \cos\phi_2 - y_2 \sin\phi_2) \sin\phi_1, \\
\dot{T}^m_{121} &= -t_{y_1} \dot{\phi}_2 \sin\phi_2 + \left(-\dot{x}_2 \cos\phi_2 + x_2 \dot{\phi}_2 \sin\phi_2 - \dot{y}_2 \sin\phi_2 - y_2 \dot{\phi}_2 \cos\phi_2\right) \sin\phi_1 \\
&= -t_{y_1} \omega \sin\phi_2 + (\upsilon \sin\phi_2 \cos\phi_2 + x_2 \omega \sin\phi_2 - \upsilon \cos\phi_2 \sin\phi_2 - y_2 \omega \cos\phi_2) \sin\phi_1 \\
&= \omega \left(-t_{y_1} \sin\phi_2 + t_{y_2} \sin\phi_1\right) = -T^m_{111} \omega.
\end{aligned}$$

By normalizing, the result is $\dot{T}_{121} = -T_{111}\omega$. Thus, the normalized dynamic system is the following:

$$\begin{aligned}
\dot{T}_{111} &= \frac{\sin\phi_1}{T_N^m}\upsilon + T_{121}\omega, & \dot{T}_{211} &= \frac{\cos\phi_1}{T_N^m}\upsilon + T_{221}\omega, \\
\dot{T}_{112} &= -\frac{\cos\phi_1}{T_N^m}\upsilon + T_{122}\omega, & \dot{T}_{212} &= \frac{\sin\phi_1}{T_N^m}\upsilon + T_{222}\omega, \\
\dot{T}_{121} &= -T_{111}\omega, & \dot{T}_{221} &= -T_{211}\omega, \\
\dot{T}_{122} &= -T_{112}\omega, & \dot{T}_{222} &= -T_{212}\omega.
\end{aligned} \tag{3.5}$$

It is worth noting that in (3.5) there are four elements that are independent on the translational velocity (T_{121}, T_{122}, T_{221} and T_{222}). It means that a change in υ does not produce a variation in these tensor elements and consequently, only orientation correction can be performed using such elements. Moreover, the normalizing factor is a kind of gain for the translational velocity.

3.2.3 Selecting Suited Outputs

The problem of taking three variables to desired values $(t_{x_2}, t_{y_2}, \sin\phi_2) = (0,0,0)$ may be completely solved with at least three outputs being controlled. However, it is also possible to find two outputs to take two variables to their desired values and then a third one is left as a degree of freedom to be corrected a posteriori. In this sense, only two outputs can be used, because defining more than two generates a non-square dynamic system, in which its non-invertibility makes difficult to prove stability of the control system.

Under the definition of a global frame in the target view, we can define the longitudinal error as the y coordinate and the lateral error as the x robot position. By taking into account three premises: 1) the values of the tensor elements in the final location, 2) the solution of the homogeneous linear system generated when the outputs are equal to zero, 3) the invertibility of the matrix relating the output dynamics with the inputs, we can state:

- It is possible to design a square control system which can correct orientation and longitudinal error. However, it leaves the lateral error as a degree of freedom. This error cannot be corrected later considering the non-holonomic constraint of the robot. Thus, this case does not have practical interest.
- It is not possible to design a square control system which allows us to correct orientation and lateral error, leaving the longitudinal error as a degree of freedom.
- It is feasible to design a square control system which can correct both longitudinal and lateral error, leaving the orientation as a degree of freedom. The orientation error can be corrected in a second step considering that the robot uses a differential drive. We concentrate in exploiting this option.

3.3 1D Trifocal Tensor-Based Control Law Design

We present the development of a two-step control law, which firstly drives the robot to a desired position and then corrects its orientation. The first step is based on solving a tracking problem for a nonlinear system in order to correct x and y positions. The second step uses direct feedback from one element of the tensor to correct orientation.

3.3.1 First Step – Position Correction

The initial location of the robot is (x_1, y_1, ϕ_1), the target location $(x_3, y_3, \phi_3) = (0,0,0)$ and the current location (x_2, y_2, ϕ_2), which varies as the robot moves. The goal is to drive the robot to the target location, i.e., to reach $(x_2, y_2, \phi_2) = (0,0,0)$. Now we define the control objective as a function of the 1D trifocal tensor elements. When the robot reaches the target, it achieves the condition given in (3.4) and therefore, the following sum of normalized tensor elements are selected as outputs:

$$\begin{aligned} \xi_1 &= T_{112} + T_{121}, \\ \xi_2 &= T_{212} + T_{221}. \end{aligned} \tag{3.6}$$

We can see that these outputs go to zero as the robot moves to the target. When $\xi_1 = 0$ and $\xi_2 = 0$ the following homogeneous linear system is given:

$$\begin{bmatrix} T_{112} + T_{121} \\ T_{212} + T_{221} \end{bmatrix} = \begin{bmatrix} \sin\phi_1 & \cos\phi_1 \\ \cos\phi_1 & -\sin\phi_1 \end{bmatrix} \begin{bmatrix} t_{x_2} \\ t_{y_2} \end{bmatrix} = \begin{bmatrix} 0 \\ 0 \end{bmatrix}.$$

This system has unique solution $t_{x_2} = 0$, $t_{y_2} = 0$ for any value of ϕ_1 ($\det(\cdot) = -1$). Thus, $(t_{x_2}, t_{y_2}, \sin\phi_2) = (0, 0, \sin\phi_2)$ is accomplished, which ensures position correction ($x_2 = 0$, $y_2 = 0$). A robust tracking controller is able to take the value of both outputs to zero in a smooth way. Let us define the tracking errors as $e_1 = \xi_1 - \xi_1^d$ and $e_2 = \xi_2 - \xi_2^d$. Thus, the error system is given as

$$\begin{bmatrix} \dot{e}_1 \\ \dot{e}_2 \end{bmatrix} = \begin{bmatrix} -\dfrac{\cos\phi_1}{T_N^m} & T_{122} - T_{111} \\ -\dfrac{\sin\phi_1}{T_N^m} & T_{222} - T_{211} \end{bmatrix} \begin{bmatrix} \upsilon \\ \omega \end{bmatrix} - \begin{bmatrix} \dot{\xi}_1^d \\ \dot{\xi}_2^d \end{bmatrix}. \tag{3.7}$$

This system has the form $\dot{\mathbf{e}} = \mathbf{M}(\mathbf{T}, \phi_1)\mathbf{u} - \dot{\xi}^d$, where $\mathbf{M}(\mathbf{T}, \phi_1)$ corresponds to the decoupling matrix and $\dot{\xi}^d$ represents a known disturbance. We need to invert the system in order to assign the desired dynamics using the inverse matrix

$$\mathbf{M}^{-1}(\mathbf{T}, \phi_1) = \frac{1}{\det(\mathbf{M})} \begin{bmatrix} T_{222} - T_{211} & T_{111} - T_{122} \\ \dfrac{\sin\phi_1}{T_N^m} & -\dfrac{\cos\phi_1}{T_N^m} \end{bmatrix}, \tag{3.8}$$

where $\det(\mathbf{M}) = \frac{1}{T_N^m}\left[(T_{122}-T_{111})\sin\phi_1 + (T_{211}-T_{222})\cos\phi_1\right]$ and $T_N^m = T_{121}^m$. At the final location $T_{221} = -\alpha t_{x_1}, T_{212} = \alpha t_{x_1}, T_{121} = \alpha t_{y_1}, T_{112} = -\alpha t_{y_1}$, where α is an unknown scale factor, and the other tensor elements are zero. The normalizing factor is never zero in our framework as described in section 3.2, although $\det(\mathbf{M}) = 0$ at the final location. This entails the problem that the rotational velocity (ω) increases to infinite as the robot reaches the target. We face this problem by commuting to a bounded control law, as described later.

We treat the tracking problem as the stabilization of the error system (3.7). We present a robust control law which solves the tracking problem using sliding mode control, which provides good properties to the control system (see appendix A.2 and [138] for details). A common way to define sliding surfaces in an error system is to take directly the errors as sliding surfaces, so that,

$$\mathbf{s} = \begin{bmatrix} s_1 \\ s_2 \end{bmatrix} = \begin{bmatrix} e_1 \\ e_2 \end{bmatrix} = \begin{bmatrix} \xi_1 - \xi_1^d \\ \xi_2 - \xi_2^d \end{bmatrix},$$

in such a way that, if there exist switched feedback gains that make the states to evolve in $\mathbf{s} = 0$, then the tracking problem is solved. We use the sliding surfaces and the *equivalent control method* in order to find switched feedback gains to drive the state trajectory to $\mathbf{s} = 0$ and maintaining it there for future time. From the equation $\dot{\mathbf{s}} = 0$, the so-called equivalent control is

$$\mathbf{u}_{eq} = \mathbf{M}^{-1}\dot{\xi}^d.$$

A control law that ensures global stabilization of the error system has the form $\mathbf{u}_{sm} = \mathbf{u}_{eq} + \mathbf{u}_{disc}$, where $\mathbf{u}_{disc}$ is a two-dimensional vector containing switched feedback gains of the form

$$\mathbf{u}_{disc} = \mathbf{M}^{-1}\begin{bmatrix} -\kappa_1 \operatorname{sign}(s_1) \\ -\kappa_2 \operatorname{sign}(s_2) \end{bmatrix},$$

where $\kappa_1 > 0$ and $\kappa_2 > 0$ are control gains. Although $\mathbf{u}_{sm}$ can achieve global stabilization of the error system, high gains may be needed, which can cause undesirable effects in real situations. We add a pole placement term in the control law to alleviate this problem

$$\mathbf{u}_{pp} = \mathbf{M}^{-1}\begin{bmatrix} -\lambda_1 & 0 \\ 0 & -\lambda_2 \end{bmatrix}\begin{bmatrix} s_1 \\ s_2 \end{bmatrix},$$

where $\lambda_1 > 0$ and $\lambda_2 > 0$ are control gains. Finally, a decoupling-based control law that achieves robust global stabilization of the system (3.7) is as follows:

$$\mathbf{u}_{db} = \begin{bmatrix} \upsilon_{db} \\ \omega_{db} \end{bmatrix} = \mathbf{u}_{eq} + \mathbf{u}_{disc} + \mathbf{u}_{pp} = \mathbf{M}^{-1}\begin{bmatrix} u_1 \\ u_2 \end{bmatrix}, \tag{3.9}$$

where $u_1 = \dot{\xi}_1^d - \kappa_1 \operatorname{sign}(s_1) - \lambda_1 s_1$, and $u_2 = \dot{\xi}_2^d - \kappa_2 \operatorname{sign}(s_2) - \lambda_2 s_2$. Note that this control law depends on the orientation of the fixed auxiliary camera ϕ_1. This orientation has to be estimated only in the initial location and can be obtained from the epipoles that relate the initial and target images. Any uncertainty in the estimation of the initial orientation can be overcome given the robustness properties of our control law, which justify the application of sliding mode control. Moreover, ϕ_1 can be fixed to zero as shown in Table 3.1 of section 3.5.1. Additionally, the sliding mode control provides robustness against the assumption of constant normalizing factor, whose effects as matched disturbances are rejected in the error system.

3.3.1.1 Solving the Singularity

We use the inverse of the decoupling matrix (3.8) to compute the control inputs, which causes a singularity problem at the final condition. The singularity affects the computation of both velocities, however υ tends to zero as the robot reaches the target. To keep ω bounded and the outputs tracking their references, the control scheme commutes to a direct sliding mode controller when $\det(\mathbf{M})$ is near to zero. This kind of controller has been studied for output tracking through singularities with good performance [67]. For this case, a bounded sliding mode controller is as follows:

$$\mathbf{u}_b = \begin{bmatrix} \upsilon_b \\ \omega_b \end{bmatrix} = \begin{bmatrix} k_\upsilon \operatorname{sign}(s_1) \\ -k_\omega \operatorname{sign}(s_2\, g(\mathbf{T})) \end{bmatrix}, \tag{3.10}$$

where k_υ and k_ω are suitable gains, and $g(\mathbf{T})$ will be defined in the stability analysis (section 3.4). It is found by achieving the negativeness of a Lyapunov function derivative. The control law (3.10) locally stabilizes the system (3.7) and is always bounded.

3.3.1.2 Desired Trajectories

The goal of the reference tracking is to take the outputs to zero in a smooth way in such a way that the robot performs a smooth motion in a desired time. To achieve that, suitable references are given by:

$$\begin{aligned} \xi_1^d &= \frac{T_{112}^{ini} + T_{121}^{ini}}{2}\left(1 + \cos\left(\frac{\pi}{\tau}t\right)\right), \quad && 0 \le t \le \tau, \\ \xi_1^d &= 0, && t > \tau, \\ \xi_2^d &= \frac{T_{212}^{ini} + T_{221}^{ini}}{2}\left(1 + \cos\left(\frac{\pi}{\tau}t\right)\right), && 0 \le t \le \tau, \\ \xi_2^d &= 0, && t > \tau, \end{aligned} \tag{3.11}$$

where τ is the time to reach the target and T_{ijk}^{ini} are the values of the tensor elements at $t = 0$. The choice of these trajectories obeys just to the requirement of a smooth zeroing of the outputs along a fixed temporal horizon. Indeed, a parabolic function may be used without difference in the resulting behavior. By defining τ, we fix the duration of the first part of the control and the time to commute to correct orientation. Note that, although initially the current image is the same than the starting one, there is enough information in the 1D trifocal tensor (3.3) to have well defined references.

3.3.2 Second Step – Orientation Correction

Once position correction has been reached in $t = \tau$, we can use any single tensor element whose dynamics depends on ω and with desired final value zero to correct orientation. We select the dynamics $\dot{T}_{122} = -T_{112}\omega$. A suitable input ω that yields T_{122} exponentially stable is

$$\omega = \lambda_\omega \frac{T_{122}}{T_{112}}, \qquad t > \tau, \tag{3.12}$$

where $\lambda_\omega > 0$ is a control gain. This rotational velocity assigns exponentially stable dynamics to T_{122}

$$\dot{T}_{122} = -T_{112}\left(k_\omega \frac{T_{122}}{T_{112}}\right) = -\lambda_\omega T_{122}. \tag{3.13}$$

Note that (3.12) never becomes singular because $T_{112} = -t_{y_1}\cos\phi_2$ for $t = \tau$ and it tends to $-t_{y_1} \neq 0$ as final value. Although only a rotation is carried out in this second step, we keep the translational velocity υ_b given in (3.10) in order to have closed loop control along the whole motion.

3.4 Stability Analysis

The control action in the first step is based on zeroing the defined outputs. So, when these outputs reach zero, the so-called *zero dynamics* in the robot system is achieved. In the particular case of the robot system (1.2) with $\ell = 0$ and output vector (3.6), this set is given as

$$Z^* = \left\{(x_2, y_2, \phi_2)^T \mid \xi_1 \equiv 0,\ \xi_2 \equiv 0\right\} = \left\{(0, 0, \phi_2)^T,\ \phi_2 \in \mathbb{R}\right\}.$$

Zero dynamics in this control system means that, when the chosen outputs are zero, the x and y-coordinates of the robot are corrected, but orientation may be different to zero. This zero dynamics yields $T_{122} = t_{y_1}\sin\phi_2$ and, therefore, when we make $T_{122} = 0$ then $\phi_2 = n\pi$ with $n \in \mathbb{Z}$, and the orientation is corrected. It is clear the exponential stability of T_{122} in the second step (3.13) for any $\lambda_\omega > 0$, and we focus on proving stability for the tracking control law.

Proposition 3.1. *Global stabilization of the system (3.7) is achieved with a commuted control law applied for $t \leq \tau$, which starts with the decoupling-based control (3.9) and commutes to the bounded control (3.10) if $|\det(\mathbf{M}(\mathbf{T}, \phi_1))| < T_h$, where T_h is a suitable threshold value.*

Proof. As mentioned above, the commutation between the decoupling-based control to the bounded one happens only when the robot is near to the target location.

For a sliding mode controller we have to prove the existence of sliding modes. This means to develop a stability proof to know if the sliding surfaces can be reached in a finite time and the state trajectory can be maintained there. Let us use the natural Lyapunov function for a sliding mode controller

$$V = V_1 + V_2,\ V_1 = \frac{1}{2}s_1^2,\ V_2 = \frac{1}{2}s_2^2, \tag{3.14}$$

which accomplishes $V(s_1 = 0,\ s_2 = 0) = 0$ and $V > 0$ for all $s_1 \neq 0,\ s_2 \neq 0$. The time-derivative of this candidate Lyapunov function is

$$\dot{V} = \dot{V}_1 + \dot{V}_2 = s_1\dot{s}_1 + s_2\dot{s}_2. \tag{3.15}$$

Now, we analyze each term of (3.15) for the decoupling based controller (3.9). After some simple mathematical simplifications we have

$$\begin{aligned}
\dot{V}_1 &= s_1\left(u_1 - \dot{\xi}_1^d\right) = s_1\left(\dot{\xi}_1^d - \kappa_1 \mathrm{sign}(s_1) - \lambda_1 s_1 - \dot{\xi}_1^d\right) = -\kappa_1 |s_1| - \lambda_1 s_1^2,\\
\dot{V}_2 &= s_2\left(u_2 - \dot{\xi}_2^d\right) = s_2\left(\dot{\xi}_2^d - \kappa_2 \mathrm{sign}(s_2) - \lambda_2 s_2 - \dot{\xi}_2^d\right) = -\kappa_2 |s_2| - \lambda_2 s_2^2.
\end{aligned}$$

$\dot{V}_1$ and $\dot{V}_2$ are negative definite if and only if the following inequalities are guaranteed for all $s_1 \neq 0,\ s_2 \neq 0$:

$$\kappa_1 > 0,\ \lambda_1 \geq 0,\ \kappa_2 > 0,\ \lambda_2 \geq 0. \tag{3.16}$$

Therefore, $\dot{V} < 0$ if and only if both inequalities (3.16) are fulfilled. So, global convergence to the sliding surfaces is achieved.

Now, let us develop the existence conditions of sliding modes for the bounded controller (3.10). The same Lyapunov function (3.14) is used, and for each term of (3.15) we have

$$\begin{aligned}
\dot{V}_1 &= -\frac{k_\upsilon \cos\phi_1}{T_N^m}|s_1| + s_1\left((T_{122} - T_{111})(-k_\omega\ \mathrm{sign}(s_2\ g(\mathbf{T}))) - \dot{\xi}_1^d\right),\\
\dot{V}_2 &= s_2\left(-\frac{k_\upsilon \sin\phi_1}{T_N^m}\ \mathrm{sign}(s_1) - \dot{\xi}_2^d\right) - k_\omega\ |s_2|\ (T_{222} - T_{211})\,\mathrm{sign}(g(\mathbf{T})).
\end{aligned}$$

Let us define $A = -k_\omega(T_{122} - T_{111})\ \mathrm{sign}(s_2\ g(\mathbf{T})) - \dot{\xi}_1^d$ and $B = -\frac{k_\upsilon \sin\phi_1}{T_N^m}$ $\mathrm{sign}(s_1) - \dot{\xi}_2^d$. In order to enforce negativeness of $\dot{V}_2$ for some value of k_ω, the function $g(\mathbf{T})$ has to be $g(\mathbf{T}) = T_{222} - T_{211}$. Hence, we have

$$\begin{aligned}
\dot{V}_1 &= -\frac{k_\upsilon \cos\phi_1}{T_N^m}|s_1| + s_1 A,\\
\dot{V}_2 &= -k_\omega\ |s_2|\ |T_{222} - T_{211}| + s_2 B.
\end{aligned}$$

We can see that

$$\dot{V}_1 \leq -\left(\frac{k_{\upsilon}\cos\phi_1}{T_N^m} - |A|\right)|s_1|,$$
$$\dot{V}_2 \leq -\left(k_{\omega}\,|T_{222} - T_{211}| - |B|\right)|s_2|.$$

$\dot{V}_1$ and $\dot{V}_2$ are negative definite if and only if the following inequalities are assured for all $s_1 \neq 0$, $s_2 \neq 0$:

$$k_{\upsilon} > \frac{T_N^m\,|A|}{\cos\phi_1}, \quad k_{\omega} > \frac{|B|}{|T_{222} - T_{211}|}. \tag{3.17}$$

Recall that sliding mode control drives the system around the sliding surface without maintaining the system exactly on the surface, which could only happen in ideal conditions of modeling [138]. Thus, the denominator of the right hand side of the last inequality does not become null for real behavior of sliding mode control. Moreover, the difference $g(\mathbf{T}) = T_{222} - T_{211}$ is the same as s_1 when the robot is reaching the target, so that, the proof of asymptotic stability with finite convergence time excludes the occurrence of $g = 0$ when $s_1 \neq 0$, as mention above (3.17). Therefore, $\dot{V} < 0$ if and only if both inequalities (3.17) are fulfilled. The bounded controller does not need any information of system parameters and thus, its robustness is implicit. □

According to the existence conditions of sliding modes, the bounded controller (3.10) is able to locally stabilize the system (3.7). Its attraction region is bigger as long as the control gains k_{υ} and k_{ω} are higher. Because of the bounded control law is also a switching one, the commutation from the decoupling-based control to the bounded one does not affect the stability of the closed loop system. The first controller ensures entering to the attraction region of the second one. Once the sliding surfaces are reached for any case of control law, the system's behavior is independent of matched uncertainties and disturbances [138]. Uncertainties in the system (3.7) due to ϕ_1 fulfill the so-called *matching condition*, and as a result, robustness of the control system is accomplished.

3.5 Experimental Evaluation

The performance of the visual control scheme of section 3.3 is evaluated through simulations and real-world experiments in this section. The results show that the main objective of driving the robot to a desired pose ((0,0,0^o) in all the cases) is attained from the chosen image measurements and even with noise in the images.

3.5.1 Simulation Results

In this section, we present some simulations of the overall control system as established in the Proposition 3.1 for the first step, and using ω (3.12) and υ_b (3.10) for the second one. Simulations have been performed in Matlab. The 1D trifocal tensor is estimated from more than five point correspondences in virtual omnidirectional images of size 1024×768. These images have been generated from a 3D scene (Fig. 3.3(a)) through the generic model for central catadioptric cameras [55]. We report results with hypercatadioptric, paracatadioptric and also fisheye cameras, which can be approximately represented with the same model [41]. Besides, the computation of the 1D trifocal tensor has been studied for fisheye cameras in [133], which supports the claim that our approach is robust to small deviations of the central camera configuration. It is worth noting that, although analytically we can deduce values of the tensor elements by substituting in (3.2) the relative location between cameras, in practice, it is troublesome when the image coordinates of two images are exactly the same. It causes that the linear estimation of the trifocal constraint degenerates for such condition. We avoid this issue by moving the robot forward for a short time before to start the control. When the robot reaches the target, there is always a minimum difference between image coordinates that is enough to prevent numeric problems to solve for the 1D trifocal tensor even in simulations. Without loss of generality, the projection center is zero for all the simulations. For the controllers, the time to reach the target position τ is fixed to 100 s, the threshold to commute to the bounded control T_h is fixed to 0.04, and the control gains are set to $\lambda_1 = 1$, $\lambda_2 = 2$, $\kappa_1 = 0.02$, $\kappa_2 = 0.02$, $\lambda_\omega = 0.3$, $k_\upsilon = 0.1$, $k_\omega = 0.05$.

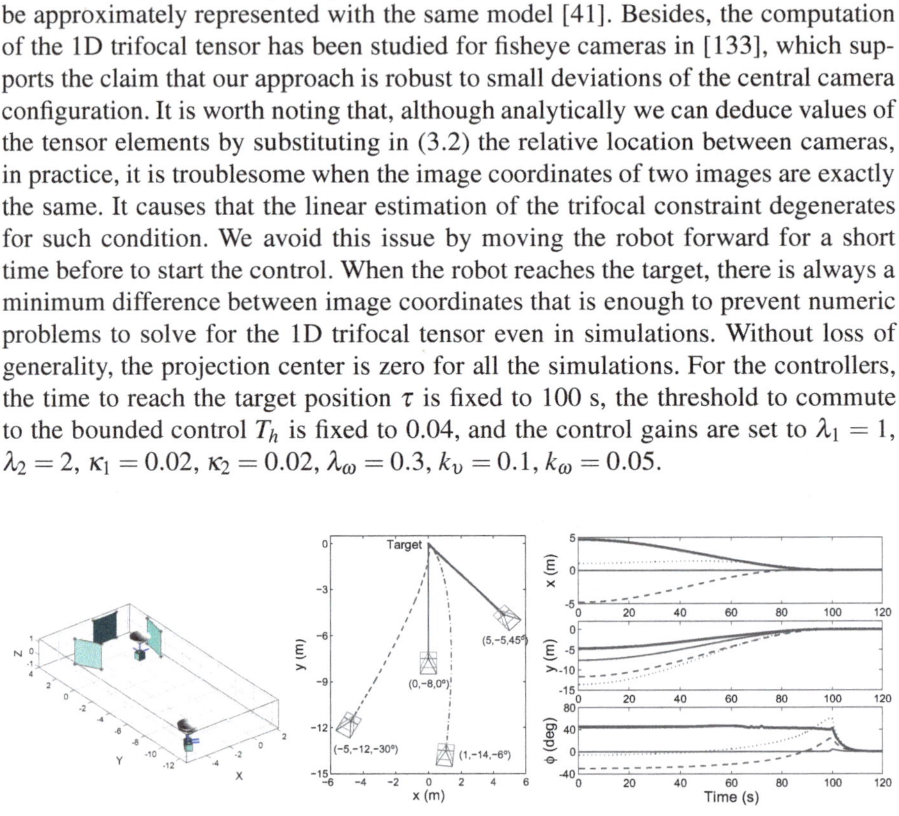

(a) (b) (c)

Fig. 3.3 Simulation results with synthetic images. (a) 3D scene. (b) Paths on the plane. (c) State variables of the robot.

Fig. 3.3(b) shows the paths traced by the robot from four different initial locations. The thick solid line starts from (5,-5,45^o), the long dashed line from (-5,-12,-30°), the solid line from (0,-8,0^o), and the short dashed line from (1,-14,-6°). In the paths, we can differentiate between three kind of autonomously performed robot motions. The solid lines correspond to a rectilinear motion to the target, while the long dashed line and the short dashed line both describe an inner curve and

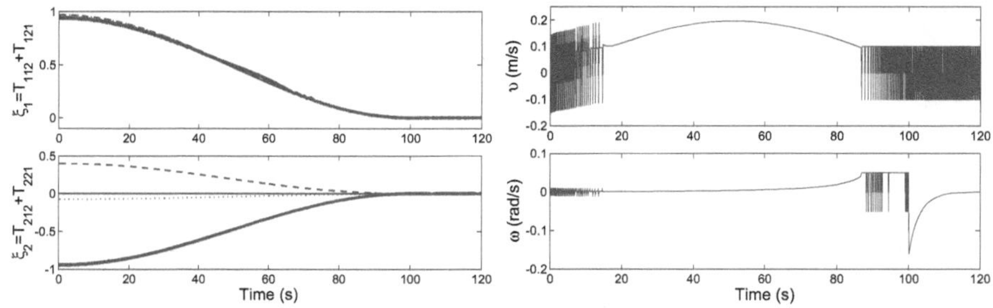

(a) (b)

Fig. 3.4 Control law performance. (a) Controlled outputs for the four cases of Fig. 3.3. (b) Example of the computed velocities for initial location (-5,-12,-30°).

an outer curve before reaching the target respectively. The rectilinear motion is obtained when the initial rotation is such that $t_{x_1} = t_{x_2} = 0$, which implies that the robot is pointing toward the target. The inner curve is generated when the initial rotation is such that $t_{x_1} = t_{x_2} > 0$ and the outer curve when the initial rotation is such that $t_{x_1} = t_{x_2} < 0$. In both later cases the robot rotation increases autonomously, and it is efficiently corrected in the second step after 100 s, as shown in Fig. 3.3(c).

We can see in Fig. 3.4(a) that both outputs are driven to zero in 100 s for all the cases. This is achieved by using bounded inputs, which are presented in Fig. 3.4(b) for the case (-5,-12,-30°). Both control inputs commute to a bounded value around 86 seconds because the determinant of the decoupling matrix falls under the fixed threshold. We can also see how the rotational velocity presents an exponential decay after 100 s, which takes the element T_{122} to zero as can be seen in Fig. 3.5. This forces the orientation to decrease with a fixed exponential rate, whose settling time is approximately 16.7 s ($5/\lambda_\omega$). This time or a threshold for T_{122} may be used to stop both of the control inputs and finish the task.

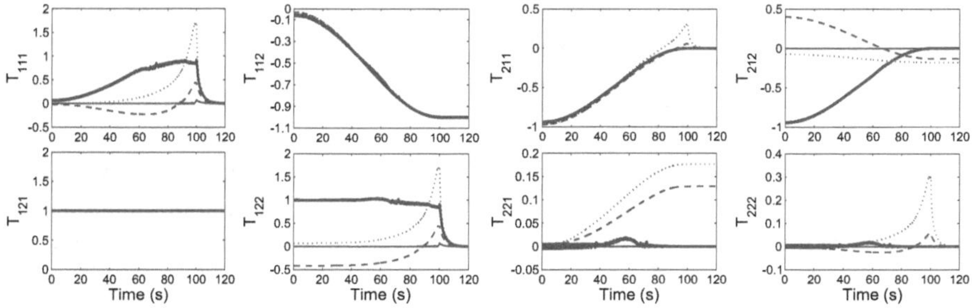

(a) (b)

Fig. 3.5 Tensor elements evolution for the four cases of Fig. 3.3. (a) Behavior of the first four elements. (b) Behavior of the second four elements.

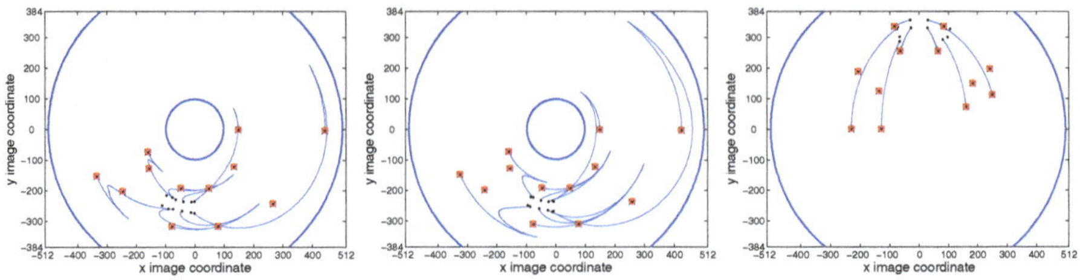

(a) (b) (c)

Fig. 3.6 Motion of the points in the image plane for three different kind of omnidirectional virtual images. (a) Hypercatadioptric. (b) Paracatadioptric. (c) Fisheye. The figures depict the point features from the initial image (represented by black points), image at the end of the motion (represented by times signs) and target image (represented by red circles).

The previous results have been obtained for three different kind of omnidirectional cameras. Fig. 3.6(a) shows the motion of the image points for the case (-5,-12,-30°), in which a hypercatadioptric camera is simulated. Fig. 3.6(b) corresponds to the case (1,-14,-6°) with a paracatadioptric camera and Fig. 3.6(c) is a fisheye camera for the initial location (0,-8,0^o).

Table 3.1 Final error for the paths in Fig. 3.3 using the control based on the trifocal tensor

	(5 m, -5 m, 45°)	(-5 m, -12 m, -30°)	(0 m, -8 m, 0°)	(1 m, -14 m, -6°)
Final error considering the initial orientation ϕ_1 as known.				
x (cm)	-0.28	0.85	0	0.91
y (cm)	0.59	0.71	0.11	-0.47
ϕ (deg)	0.10	0.02	0	0.08
Final error fixing $\phi_1 = 0$ in the controller.				
x (cm)	-0.51	0.77	0	0.98
y (cm)	0.86	0.39	0.11	-0.25
ϕ (deg)	0.11	0.01	0	0.07

Table 3.1 shows that the target location is reached with good accuracy. The results in the first part of the table are obtained considering that the initial orientation ϕ_1 is known for each case. On the other hand, the second part of the table shows that the precision is preserved even if the initial orientation is fixed to $\phi_1 = 0$ in the controller for all the cases. We can assert that similar accuracy is obtained by fixing ϕ_1 in the range $-30 \leq \phi_1 \leq 30$, since that the sliding mode control law is robust to parametric uncertainty. For all the experiments, the mean squared tracking error is very low, in the order of 1×10^{-5}.

Fig. 3.7(a) shows the good performance of the approach under image noise for initial pose (5,-10,35^o). The added noise has a standard deviation of 1 pixel and the time to reach the target (τ) is set to 60 s. The control inputs are affected directly by the noise, as can be seen in Fig. 3.7(b). Nevertheless, the outputs are regulated properly to the desired reference as shown in Fig. 3.7(c). The presence of the noise can be observed in the image points motion of Fig. 3.8(a), which results in the behavior of the tensor elements that is presented in Fig. 3.8(b)-(c).

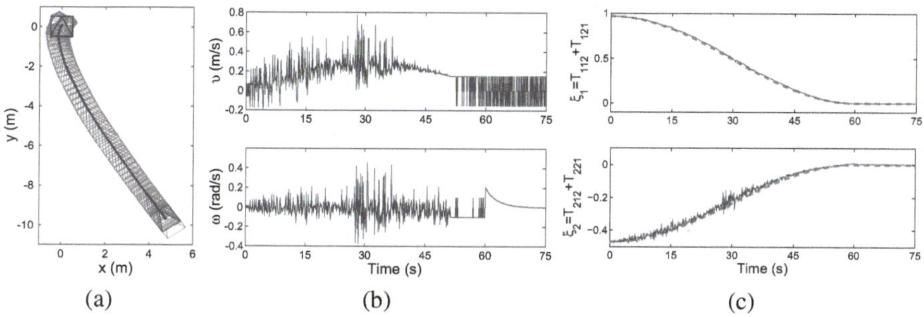

Fig. 3.7 Control law performance with image noise. (a) Resultant robot path. (b) Control inputs. (c) Controlled outputs.

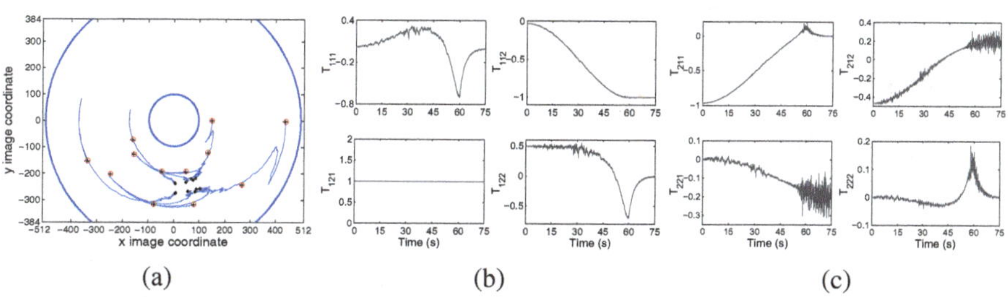

Fig. 3.8 Visual measurements from synthetic images with image noise. (a) Motion of the image points. The black points represent the initial image, the red circles represent the target image and the black plus signs represent the image at the end of the motion. (b) Behavior of the first four tensor elements. (c) Behavior of second four tensor elements.

3.5.2 Experiments with Real Data

This section describes an analysis of the behavior of the presented control scheme through experiments with omnidirectional images. Two techniques of extracting the required features are employed. The first case corresponds to the use of the well known scale invariant feature transform (SIFT) [88] and, in the second case, we use the Lucas-Kanade pyramidal algorithm [92], [1]. These experiments are performed off-line, which means that a sequence of images were taken and then used to compute the 1D trifocal tensor and the control inputs to analyze the effect of the feature extraction. This analysis is a key factor toward the real-world experimentation of the next section. Sliding mode control requires a relatively high closed loop frequency, around 10 Hz as minimum, and consequently, the computational cost of the feature extraction and the matching process becomes very important. For these experiments, we use an omnidirectional system with a camera Sony XCD-X7101CR and a mirror Neovision H3S (Fig. 1.2(b)) to capture images of size 1024×768. The image data is acquired using the free software tool Player. The commanded robot motion is a slight curve going forward and finishing with a rotation.

An important parameter required to obtain the bearing measurements is the projection center. We have tested the singleness of this point in our imaging system by estimating the projection center along a sequence. Like in [121], the center is robustly estimated using a random sample consensus (RANSAC) approach from 3D vertical lines, which project in radial lines for central imaging systems. Results have shown that our imaging system properly approximates a single view point configuration, with standard deviation of around 1 pixel for each image coordinate of the estimated center. For the size of images that we are using, these deviations have a negligible effect in the computation of bearing measurements and thus, we have fixed the projection center to ($x_0 = 541$, $y_0 = 405$).

3.5.2.1 Behavior Using SIFT Features

We have implemented a 1D trifocal tensor estimation algorithm by solving the trifocal constraint for at least five points that are extracted using SIFT and robustly matched using RANSAC. Fig. 3.9 shows an example of the SIFT [88] point matches (34 good matches) used to compute the tensor. The five-point method reduces the number of iterations required for the robust estimation, however, the computation time of the 1D trifocal tensor with this method is still very high (approximately 5 seconds per iteration).

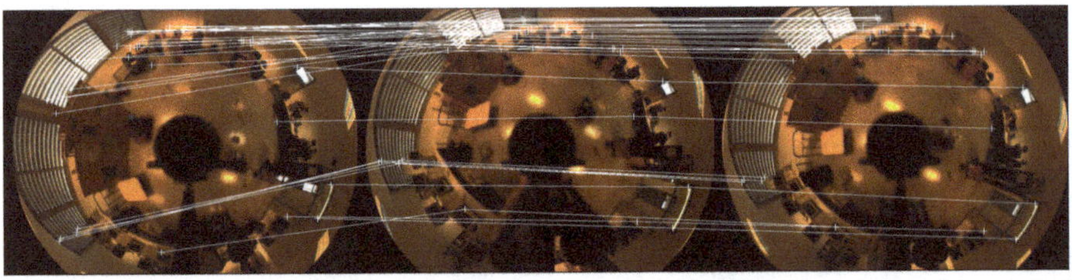

Fig. 3.9 Robust SIFT matching between three omnidirectional images with translation and rotation between them. The lines between images show 34 corresponding features, which have been extracted using SIFT and matched robustly to be the entries of the 1D trifocal tensor estimation.

Moreover, as can be seen in Fig. 3.10, the 1D trifocal tensor estimation is very unstable even having correct matches. It happens because in some cases the matches are concentrated in a region of the image. Besides, due to the property of SIFT features of being a region in the image, the effective coordinates of the features may change discontinuously along the sequence. We can see that the elements of T_2 are the most unstable, in particular when the current image is close to the target image (around 35 seconds), however, after this time the first control step is finishing and the noisy elements are not used anymore.

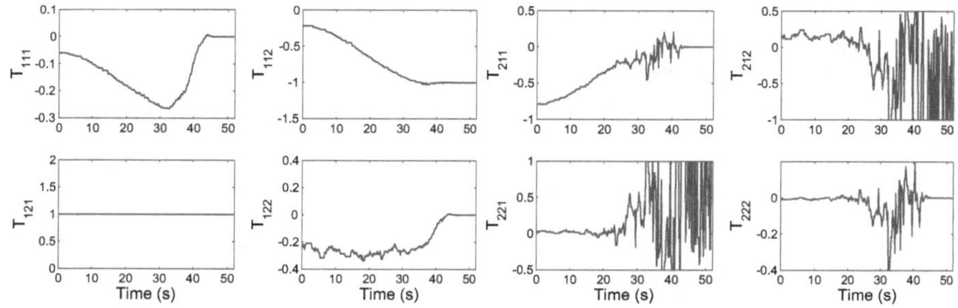

(a) (b)

Fig. 3.10 Performance of the 1D trifocal tensor estimation using SIFT features. (a) Behavior of the first four tensor elements. (b) Behavior of the second four tensor elements.

Fig. 3.11(a) shows how the reference trajectory for the first output is well approximated while output two is not close to its reference. Fig. 3.11(b) presents the computed control inputs. The translational velocity approximately describes the forward motion; however, the rotational velocity is very noisy.

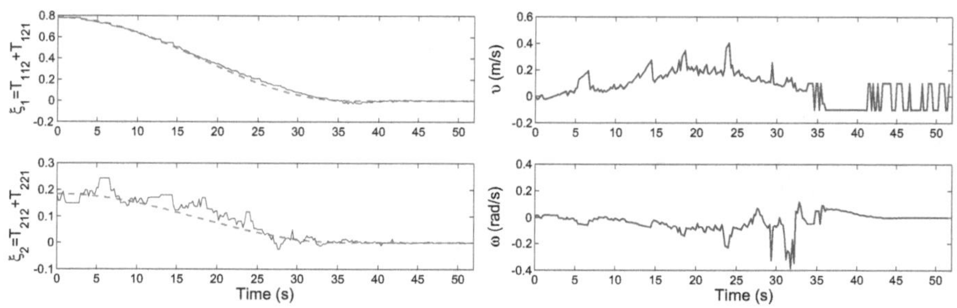

(a) (b)

Fig. 3.11 Behavior of the control law using SIFT features. (a) Outputs and their references. (b) Computed velocities.

3.5.2.2 Behavior Using Tracking of Features

In order to achieve an adequate closed loop frequency, we evaluate the strategy of tracking a set of chosen points using the Lucas-Kanade algorithm [92], [1]. The tracking of features has been extensively applied for visual servoing purposes [97]. It allows us to have the matching between features for each iteration without additional computations, which makes the scheme feasible for real-world experimentation. Additionally, the smooth motion of the image features with the Lucas-Kanade tracker results in a stable tensor estimation. We have defined 12 point features to be tracked along the same image sequence and then, the corresponding point coordinates are used to estimate the 1D trifocal tensor and the velocities as given for our control law. The resulting behavior of the trifocal tensor elements presented in Fig. 3.12 shows that they are more stable than in the case of SIFT features. However, a similar behavior is obtained at the end for the elements T_{212} and T_{221}. According to Fig. 3.13(a) both of the outputs are close to their reference trajectories, and consequently, the computed velocities in Fig. 3.13(b) actually describe the real motion of the camera.

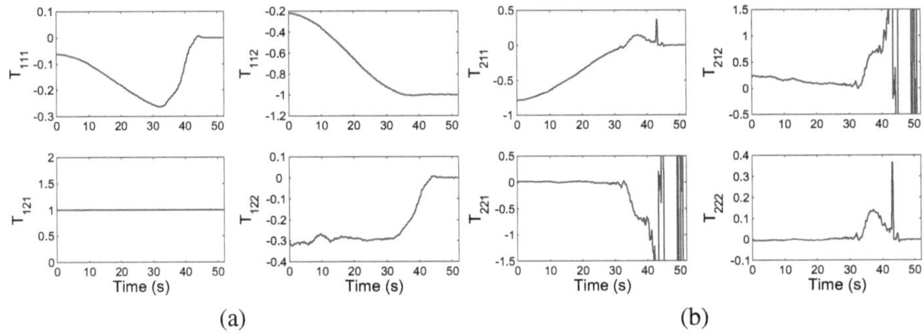

Fig. 3.12 Performance of the 1D trifocal tensor estimation using tracking of point features. (a) Behavior of the first four tensor elements. (b) Behavior of the second four tensor elements

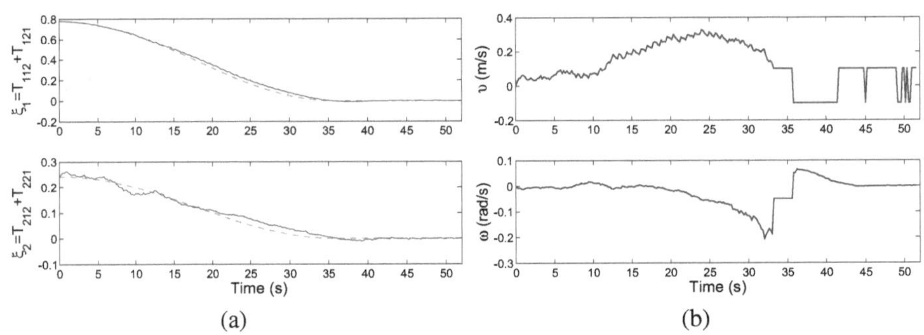

Fig. 3.13 Behavior of the control law using tracking of point features. (a) Outputs and their references. (b) Computed velocities.

3.5.3 Real-World Experiments

The complete control approach has been tested in closed loop with real conditions using the robot Pioneer 3-AT from Adept Mobile Robots as shown in Fig. 1.1(b). The same hypercatadioptric imaging system of the previous section is used, but now the images are acquired at a size of 800×600 pixels. The projection center has

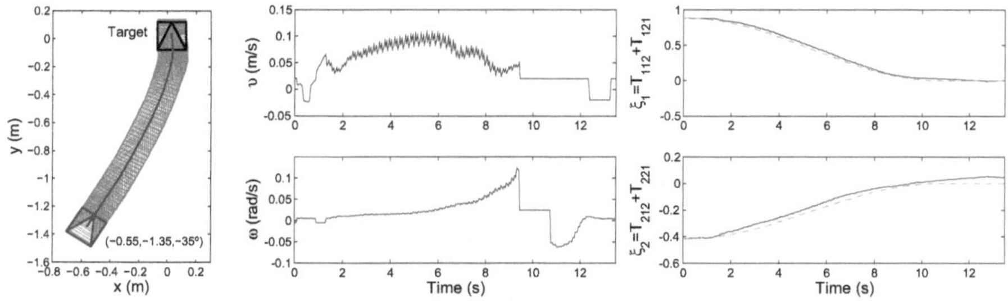

(a) (b) (c)

Fig. 3.14 Experimental results with the control law in closed loop. (a) Resultant path. (b) Computed velocities. (c) Controlled outputs. The data to plot the path is given by the robot odometry.

been fixed according to a calibration process to ($x_0 = 404$, $y_0 = 316$). The observed scene has been set up with features on three different planes in order to ensure a sufficient number of points in the scene. However, points not belonging to these planes are also used to achieve a total of 15 points, which are manually matched in the three initial images. We have implemented these experiments using the tracking of features because its low computational cost. It gives good closed loop frequency, which leads to a good behavior in the 1D trifocal tensor estimation, as described in the previous section.

Fig. 3.14(a) presents the resultant path, given by odometry, of the closed loop control from the initial location (-0.55 m,-1.35 m,-35°) for one of the experimental runs. The duration of the task is almost 14 s, the final position error is around 2 cm and the orientation error is practically negligible. The time τ for the execution of the first step is set to 9.4 s through fixing a number of iterations in our control software. Before that, we can see in Fig. 3.14(b) that the bounded sliding mode control law is applied due to the singularity of the decoupling-based controller. Fig. 3.14(c) shows that the behavior of the outputs is always close to the desired one but with a small error. The reason of the remaining error is that our robotic platform is not able to execute commands at a frequency higher than 10 Hz, and consequently the performance of the sliding mode control is not the optimum. However, the experimental evaluation shows that the closed loop frequency of 10 Hz is enough to obtain small chattering effect. As long as the frequency could be increased the chattering could be even smaller. According to Fig. 3.15(a) the motion of the image points along the sequence does not exhibit a jeopardizing noise, in such a way that the tensor elements evolve smoothly during the task, as presented in Fig. 3.15(b)-(c). Fig. 3.16 shows a sequence of some images taken by the robot camera.

In accordance to the results and the methodology presented along this chapter, we can state that the main advantages of using the 1D trifocal tensor on visual servoing are that the geometric constraint improves the robustness to image noise by filtering the data, allows applying the control approach with any visual sensor obeying approximately a central projection model and avoids the problem of short

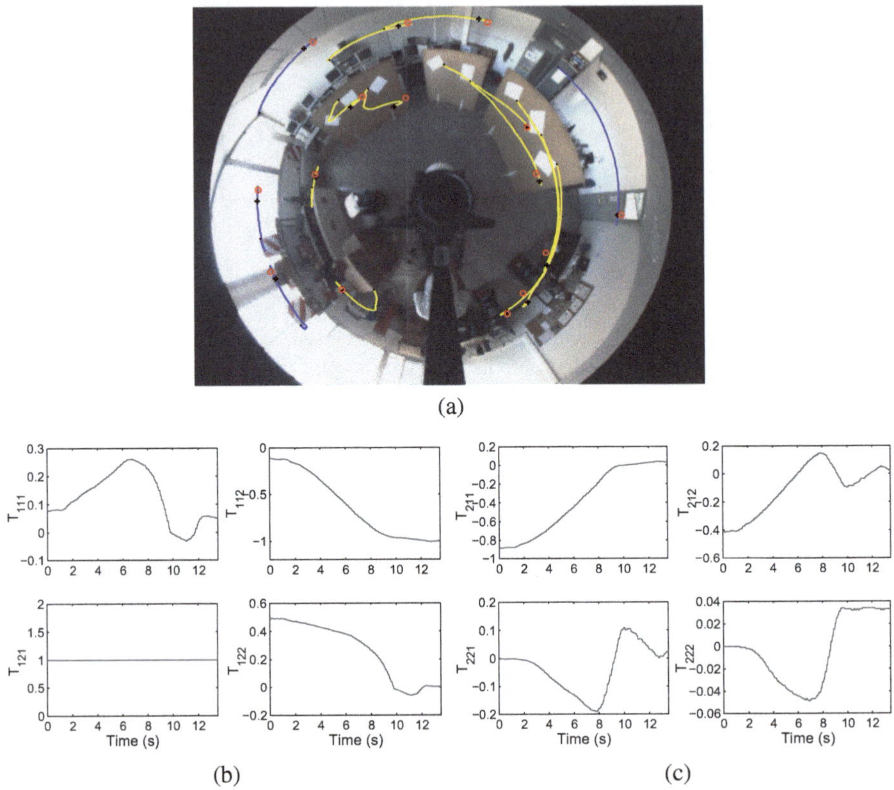

Fig. 3.15 Behavior of the visual measurements for the real experiments. (a) Motion of the image points. (b) Evolution of the first four tensor elements. (c) Evolution of the second four tensor elements.

baseline by exploiting the information of three views. Thus, total correction of both position and orientation is ensured without commuting to any visual constraint other than the 1D trifocal tensor. Since we assume planar motion and the omnidirectional camera is placed looking upward, the use of the 1D trifocal tensor particularly adapts to the property of this imaging process of preserving bearing information. Therefore, we have achieved an algorithm that is independent of the radial distortion induced by lenses and mirrors.

From a control theory point of view, an additional advantage of our approach with respect to the basic image-based visual servoing schemes is that the selected outputs allow us to prove stability on the basis of a square control system. Additionally, we have incorporated robustness properties to the closed loop by using sliding mode control. The lack of clear stability properties and robustness has been a serious concern for image-based visual servoing schemes [31]. However, the cost of proving stability with large range of convergence and without local minima, is to be

Fig. 3.16 Sequence of some of the omnidirectional images taken from the hypercatadioptric robot camera during the real experiments. The first image at the top of the figure is the target image and the next to the right is the initial image. The last image at the bottom of the figure is the image at the end of the motion. This last image is basically the same as the target one as expected.

limited to the application of our approach on differential drive robots, for which the final remaining orientation can be corrected.

In the evaluation of the detailed visual control scheme we have not considered explicitly maneuvers to be carried out. Thus, only initial locations allowing a direct motion toward the target without maneuvers are considered. However, the pose regulation for some particular initial locations can be addressed by maneuvering, for instance, when there is only a lateral error. In those cases, the robot can be driven to a general configuration by defining adequate references for the outputs. Therefore, the presented approach is a time-varying two-step control law that complies with the Brockett's theorem in the sense that some maneuvers can be carried our if required.

3.6 Closure

Along this chapter we have presented a novel image-based scheme to perform visual servoing for pose regulation of mobile robots using the elements of the 1D trifocal tensor directly in the control law. This tensor is an improved visual measurement with respect to the epipolar geometry, more general, more robust and without the drawbacks of the epipolar geometry, like the problem of short baseline. Moreover, it allows exploiting the property of omnidirectional images of preserving bearing information in a natural way. The presented visual control utilizes the usual teach-by showing strategy without requiring any a priori knowledge of the scene and does not need any auxiliary image. This control scheme is valid for any visual sensor obeying the central projection model and without requiring a complete calibration. A two-step control law ensures total correction of position and orientation without need of commuting to any visual constraint other than the 1D trifocal tensor. An adequate two-dimensional error function is used in such a way that a sliding mode control law in a square system ensures stability, with large region of convergence, no local minima and robustness of the closed loop. The effectiveness of our approach is tested through simulations and real-world experiments with a hypercatadioptric camera.

Chapter 4
Dynamic Pose-Estimation for Visual Control

Abstract. In this chapter, we change from the image-based approach to estimate the position and orientation of the camera-robot system in order to regulate the pose in the Cartesian space. This provides the benefits of including a kind of *memory* in the closed loop, which reduces the dependence of the control on the visual data and facilitates the planning of complex tasks. The camera-robot pose is recovered using a dynamic estimation scheme that exploits visual measurements given by the epipolar geometry and the trifocal tensor. The interest of the chapter is a novel observability study of the pose-estimation problem from measurements given by the aforementioned geometric constraints, as well as the demonstration that the estimated pose is suitable for closed loop control. Additionally, a benefit of exploiting measurements from geometric constraints for pose-estimation is the generality of the estimation scheme, in the sense that it is valid for any visual sensor obeying a central projection model. The effectiveness of the approach is evaluated via simulations and real-world experiments.

4.1 Introduction

In control theory, the state estimation is an important tool for the implementation of full state feedback control laws. In the case of a wheeled mobile robot, the state of the system corresponds to the position and orientation with respect to a coordinate system, which can be fixed to a target location. One of the first complete schemes for automatic localization, map-maintenance and goal-directed steering is presented in [46]. The availability of the robot state may facilitate the planning of a navigation task using a visual servoing approach [52], for instance, making possible to define a subgoal location for obstacle avoidance. Herein, we rely on the use of state estimation in visual control to reduce the dependence on the visual information because the knowledge of the previous motion may be used to recover some lost data or to define a new goal location when required. In contrast to static pose-estimation approaches [24], [51], [119], [43], where the pose is extracted from the decomposition of a particular mathematical entity at each instant time, dynamic pose-estimation is

H.M. Becerra and C. Sagüés, *Visual Control of Wheeled Mobile Robots*,
Springer Tracts in Advanced Robotics 103,
DOI: 10.1007/978-3-319-05783-5_4,

an alternative that has been little exploited for visual servoing purposes of wheeled mobile robots. The pose estimated dynamically depends on its previous time-history and this process improves the robustness of the estimation by filtering and smoothing the visual measurements.

In this chapter, an efficient mapless pose-estimation scheme for application in visual servoing of wheeled mobile robots is presented. The scheme exploits dynamic estimation and uses visual measurements provided by a geometric constraint: the epipolar geometry or the 1D trifocal tensor. These geometric constraints have shown its effectiveness to recover relative camera locations in [60], [113], [49], [59] as static approaches. Additionally, the epipolar geometry and the trifocal tensor have been exploited for visual servoing of mobile robots as raw feedback information in previous approaches [87], [100], [85], as well as in our proposals of the two previous chapters, but they have never been used in dynamic estimation for visual servoing purposes. A work in the field of computer vision shows the applicability of the trifocal tensor for dynamic camera-motion estimation [143], which presents a filtering algorithm to tackle the vision-based pose-tracking problem for augmented reality applications.

The basics of the scheme presented in this chapter has been introduced in [19] and [20]. The former work presents a basic estimation scheme using the 1D trifocal tensor and the second introduces a complete analysis of the epipolar geometry for dynamic pose-estimation. Given the good benefits of the 1D trifocal tensor as measurement, the application of the estimated pose in a visual servoing scheme has been analyzed in [21]. The approach is valid for any visual sensor obeying approximately a central projection model and only requires the center of projection of the omnidirectional images or the principal point for conventional ones, therefore, a semicalibrated scheme is obtained. A novel observability analysis of the estimation scheme is developed using nonlinear and linear tools, which gives evidence about the rich information encoded in the geometric constraints. Since the control scheme uses feedback of the estimated state, a stability analysis of the closed loop shows the validity of a separation principle between estimation and control in our nonlinear framework.

The presented approach integrates the rich visual information provided by a geometric constraint into an effective estimation scheme, which results in an adequate compromise between accuracy and computational cost. Thus, Cartesian information is available to be used for different types of tasks, such as homing for large displacements, path following or reactive navigation. The innovative aspects of the approach presented in this chapter are that, to the authors' knowledge, it is the first semicalibrated pose-estimation scheme and valid for the large group of cameras with a central configuration. The approach does not need a target model, scene reconstruction or depth information. Additionally, the control scheme is able to drive the robot to a desired position and also orientation (pose regulation problem), through smooth velocities given by a single controller.

The chapter is organized as follows. Section 4.2 presents an extensive observability analysis and the estimation scheme for measurements from the epipolar geometry and the 1D trifocal tensor. Section 4.3 details the control law and the stability of the closed loop with feedback of the estimated pose. Section 4.4 shows the performance evaluation through simulations and real-world experiments using a hypercatadioptric imaging system, and finally, Section 4.5 gives a closure of the chapter.

4.2 Dynamic Pose-Estimation from a Geometric Constraint

In this chapter, we use the information provided by a minimum set of visual measurements in order to estimate the camera location of the non-holonomic camera-robot (1.4). The visual measurements are taken from a geometric constraint, and they pass through a filtering process given by a dynamic estimation scheme. This process provides robustness against image noise and reduces the dependence on data of the image space.

Given that the model of the system and the measurement model are nonlinear, the estimation problem faced herein is intrinsically nonlinear. So, it is more adequate to take the nonlinear properties of the problem into account using the appropriate theory. In this chapter, new results on the observability of the robot pose with measurements from the epipolar geometry or the 1D trifocal tensor are reported using nonlinear tools. Additionally, this analysis leads us to conclude about the effect of the control inputs over the estimation scheme.

In this section, it is shown the feasibility of estimating the robot pose through a suitable selection of a vector of measurements $h(\mathbf{x})$ (eventually scalar). We present a novel observability analysis in continuous and discrete time in order to show the benefits of the selected visual measurements.

4.2.1 Observability Analysis with the Epipoles as Measurement

It has been usual to exploit the epipolar geometry that relates two images to extract the relative rotation and translation between the corresponding camera locations in static approaches [119], [79], [43]. We use the horizontal coordinate of the epipoles to estimate the current camera-robot pose $\mathbf{C}_2 = (x, y, \phi)$ through a dynamic approach. These epipoles can be expressed as a function of the current pose as

$$\begin{aligned} e_{cur} &= \alpha_x \frac{x\cos\phi + y\sin\phi}{y\cos\phi - x\sin\phi} = \alpha_x \frac{e_{cn}}{e_{cd}}, \\ e_{tar} &= \alpha_x \frac{x}{y}. \end{aligned} \tag{4.1}$$

This section shows the feasibility of the estimation by analyzing the observability of the camera-robot state with the vector of measurements

$$h^e(\mathbf{x}) = \left[h_1^e = e_{cur} \; h_2^e = e_{tar} \right]^T, \tag{4.2}$$

where the superscript e refers to epipoles as measurements.

4.2.1.1 Nonlinear Observability Analysis

This section utilizes the theory developed in [65] and introduced in appendix A.3.1 using the epipoles as measurements to conclude about the observability of the camera-robot system. The conclusions are summarized in the following lemma.

Lemma 4.1. *The continuous camera-robot system (1.3) with both epipoles as measurements (4.2) is a locally weakly observable system. Moreover, this property is maintained even by using only the target epipole as measurement.*

Proof. The proof of this lemma is done by finding the space spanned by all possible Lie derivatives and verifying its dimension. This space is given as

$$\Omega^e = \left(h_p^e, L_{g_1}^1 h_p^e, L_{g_2}^1 h_p^e, L_{g_1}^2 h_p^e, L_{g_2}^2 h_p^e, L_{g_1 g_2} h_p^e, \ldots\right)^T \text{ for } p = 1,2.$$

First, the Lie derivatives given by the current epipole as measurement ($h_1^e = e_{cur}$) are presented. As a good approach, the search of functions in the Lie group is constrained for $n-1$, where $n = 3$ in our case.

$$L_{g_1}^1 h_1^e = \nabla h_1^e \cdot g_1 = \frac{\alpha_x}{e_{cd}^2}\left[y \;\; -x \;\; x^2+y^2\right]\begin{bmatrix}-\sin\phi \\ \cos\phi \\ 0\end{bmatrix} = -\alpha_x \frac{e_{cn}}{e_{cd}^2},$$

$$L_{g_2}^1 h_1^e = \nabla h_1^e \cdot g_2 = \frac{\alpha_x}{e_{cd}^2}\left[y \;\; -x \;\; x^2+y^2\right]\begin{bmatrix}-\ell\cos\phi \\ -\ell\sin\phi \\ 1\end{bmatrix} = \alpha_x \frac{x^2+y^2-\ell e_{cd}}{e_{cd}^2},$$

$$\begin{aligned}
L_{g_1}^2 h_1^e &= \nabla L_{g_1}^1 h_1^e \cdot g_1 = 2\alpha_x \frac{e_{cn}}{e_{cd}^3},\\
L_{g_2}^2 h_1^e &= \nabla L_{g_2}^1 h_1^e \cdot g_2 = \alpha_x \frac{e_{cn}\left(2\left(x^2+y^2\right)-3\ell e_{cd}\right)}{e_{cd}^3},\\
L_{g_1 g_2} h_1^e &= -\alpha_x \frac{2e_{cn}^2-\ell e_{cd}}{e_{cd}^3},\\
L_{g_2 g_1} h_1^e &= -\alpha_x \frac{x^2+y^2-\ell e_{cd}+e_{cn}^2}{e_{cd}^3}.
\end{aligned}$$

To verify the dimension of the space spanned by these functions, the gradient operator is applied to obtain the matrix $\mathbf{O}_{cur}$ (4.3). Recall that we use the notation $s\phi = \sin\phi$ and $c\phi = \cos\phi$. Given the complexity of the entries of the matrix, only four rows are shown, however, the matrix is of rank two even with more rows.

$$
\begin{aligned}
\mathbf{O}_{cur} &= \left[\left(\nabla h_1^e\right)^T \; \left(\nabla L_{g_1}^1 h_1^e\right)^T \; \left(\nabla L_{g_2}^1 h_1^e\right)^T \; \left(\nabla L_{g_1}^2 h_1^e\right)^T \; \cdots \right]^T \\
&= \frac{\alpha_x}{e_{cd}^2} \begin{bmatrix} y & -x & x^2+y^2 \\ -\left(y+s\phi e_{cn}\right)/e_{cd} & \left(x+c\phi e_{cn}\right)/e_{cd} & -\left(x^2+y^2+e_{cn}^2\right)/e_{cd} \\ -\left(\ell s\phi e_{cd}-2ye_{cn}\right)/e_{cd} & \left(\ell c\phi e_{cd}-2xe_{cn}\right)/e_{cd} & e_{cn}\left(2\left(x^2+y^2\right)-\ell e_{cd}\right)/e_{cd} \\ \left(2c\phi e_{cd}+6s\phi e_{cn}\right)/e_{cd}^2 & \left(2s\phi e_{cd}-6c\phi e_{cn}\right)/e_{cd}^2 & \left(2\left(yc\phi-xs\phi\right)^2+6e_{cn}^2\right)/e_{cd}^2 \\ \vdots & \vdots & \vdots \end{bmatrix}.
\end{aligned} \tag{4.3}
$$

It is required that the Lie derivatives obtained from the target epipole as measurement ($h_2^e = e_{tar}$) provide one row linearly independent on those previous in (4.3). These new Lie derivatives are the following:

$$
\begin{aligned}
L_{g_1}^1 h_2^e &= \nabla h_2^e \cdot g_1 = \frac{\alpha_x}{y^2}\left[y \; -x \; 0\right] \begin{bmatrix} -\sin\phi \\ \cos\phi \\ 0 \end{bmatrix} = -\frac{\alpha_x}{y^2} e_{cn}, \\
L_{g_2}^1 h_2^e &= \nabla h_2^e \cdot g_2 = \frac{\alpha_x}{y^2}\left[y \; -x \; 0\right] \begin{bmatrix} -\ell\cos\phi \\ -\ell\sin\phi \\ 1 \end{bmatrix} = -\frac{\alpha_x \ell}{y^2} e_{cd}, \\
L_{g_1}^2 h_2^e &= \nabla L_{g_1}^1 h_2^e \cdot g_1 = \frac{2\alpha_x}{y^3} \cos\phi \, e_{cn}, \\
L_{g_2}^2 h_2^e &= \nabla L_{g_2}^1 h_2^e \cdot g_2 = \frac{\alpha_x \ell}{y^3} \left(y e_{cn} - 2\ell \sin\phi \, e_{cd}\right).
\end{aligned}
$$

By applying the gradient operator, the matrix $\mathbf{O}_{tar}$ (4.4) is obtained, which effectively provides the additional linearly independent row to span the three dimensional state space.

$$
\begin{aligned}
\mathbf{O}_{tar} &= \left[\left(\nabla h_2^e\right)^T \; \left(\nabla L_{g_1}^1 h_2^e\right)^T \; \left(\nabla L_{g_2}^1 h_2^e\right)^T \; \left(\nabla L_{g_1}^2 h_2^e\right)^T \; \left(\nabla L_{g_2}^2 h_2^e\right)^T \right]^T \\
&= \frac{\alpha_x}{y^4} \begin{bmatrix} y^3 & -y^2 x & 0 \\ -y^2 c\phi & y\left(ys\phi+2xc\phi\right) & -y^2\left(yc\phi-xs\phi\right) \\ \ell y^2 s\phi & \ell y\left(yc\phi-2xs\phi\right) & \ell y^2\left(xc\phi+ys\phi\right) \\ 2yc\phi^2 & -2c\phi\left(3xc\phi+2ys\phi\right) & 2y\left(y\left(c\phi^2-s\phi^2\right)-2xs\phi c\phi\right) \\ -\ell y\left(yc\phi+2\ell s^2\phi\right) & o_{52} & \ell y\left(\left(y-2\ell c\phi\right)e_{cd}+2\ell s\phi e_{cn}\right) \end{bmatrix},
\end{aligned} \tag{4.4}
$$

where $o_{52} = -\ell\left(y^2\sin\phi + y\cos\phi\left(2x - 4\ell\sin\phi\right) + 6\ell x \sin^2\phi\right)$. Thus, from the matrix $\mathbf{O}^e = \left[\mathbf{O}_{cur}^T \; \mathbf{O}_{tar}^T\right]^T$, we can state that the system has the property of locally weak observability and the three state variables constituting the camera-robot pose can be estimated from these two measurements. Moreover, the matrix (4.4) is full rank by itself, which means that the rank condition A.6 is satisfied by using the target epipole as unique measurement. So, the camera-robot system (1.3) with both epipoles as measurements is locally weakly observable and this property is achieved even by using only the target epipole as measurement. □

Notice that the previous proof implicitly considers the action of both velocities, however, we can analyze the effect for each one of them. For simplicity, we observe the results with the target epipole as measurement, i.e., analyzing the matrix (4.4). On one hand, it can be shown that

$$det\left(\left[(\nabla h_2^e)^T \ (\nabla L_{g_1}^1 h_2^e)^T \ (\nabla L_{g_1}^2 h_2^e)^T\right]^T\right) = -2\alpha_x y^2 e_{cn},$$

which means that, when only a translational velocity is being applied, the matrix loses rank if the current epipole is zero. In other words, observability is lost if the robot is moving forward along the line joining the projection center of the cameras because the target epipole remains unchanged. Notice that if the robot is not in the described condition, observability is guaranteed by a translational velocity different than zero. On the other hand,

$$det\left(\left[(\nabla h_2^e)^T \ (\nabla L_{g_2}^1 h_2^e)^T \ (\nabla L_{g_2}^2 h_2^e)^T\right]^T\right) = \alpha_x \ell^2 y^2 d(\mathbf{x}), \text{ with } d(\mathbf{x}) \neq 0 \text{ for all } \mathbf{x} \neq 0.$$

This means that the rotational velocity provides observability if and only if the camera is shifted from the axis of rotation ($\ell \neq 0$), given that in this situation, the target epipole changes as the robot rotates. Thus, the control strategy should provide the appropriate excitation, at least non-null rotational velocity, in order to ensure the property of observability for any condition, even when the robot is looking directly toward the target.

4.2.1.2 Analysis as Piece-Wise Constant System

An estimation scheme can be implemented through a discrete Kalman filtering approach in order to estimate the camera-robot pose $\hat{\mathbf{x}}_k = \left[\hat{x}_k \ \hat{y}_k \ \hat{\phi}_k\right]^T$ from visual measurements. This estimation scheme provides generality to the visual servoing approach in comparison to other options like nonlinear observers, which are designed for a particular system. Additionally, a Kalman estimator allows to change easily the set of measurements during the operation. Given that we are dealing with a nonlinear estimation problem, an extended Kalman filter (EKF) is an effective way to solve it. In order to achieve a good compromise between accuracy in the estimation and computational cost, we use the basic form of the EKF as detailed in appendix A.4.

In the previous section, it is proved that the camera-robot model with the epipoles as measurement is an observable system from a nonlinear point of view. The linearization process may affect the observability and inconsistent estimation may be obtained [68]. Since the EKF is based on linearization of the system (1.5) and outputs (4.1), in this section, the observability property of the linear approximation $(\mathbf{F}_k, \mathbf{G}_k, \mathbf{H}_k^e)$ is investigated. The matrices $\mathbf{F}_k$ and $\mathbf{G}_k$ are given by:

$$\mathbf{F}_k = \frac{\partial f}{\partial \mathbf{x}_k}, \quad \mathbf{G}_k = \frac{\partial f}{\partial \mathbf{u}_k},$$

and they are detailed in A.13. The corresponding measurement matrix from the epipoles is

$$\mathbf{H}_k^e = \frac{\partial h}{\partial \mathbf{x}_k} = \begin{bmatrix} \frac{\alpha_x}{e_{cd,k}^2} \left[y_k \; -x_k \; x_k^2 + y_k^2 \right] \\ \frac{\alpha_x}{y_k^2} \left[y_k \; -x_k \; 0 \right] \end{bmatrix}, \tag{4.5}$$

where $e_{cd,k} = y_k \cos\phi_k - x_k \sin\phi_k$.

Lemma 4.2. *The linear approximation* $(\mathbf{F}_k, \mathbf{G}_k, \mathbf{H}_k^e)$ *of the discrete nonlinear system (1.5) and epipoles as measurements (4.1) used in an EKF-based estimation scheme is an observable system. Moreover, observability is achieved by using only the target epipole as measurement.*

Proof. Firstly, we verify the property of local observability, which is given by the typical observability matrix

$$\mathbf{O}_k^e = \left[(\mathbf{H}_k^e)^T \; (\mathbf{H}_k^e \mathbf{F}_k)^T \; \cdots \; (\mathbf{H}_k^e \mathbf{F}_k^{n-1})^T \right]^T.$$

This is a 6×3 matrix that can be built by stacking the local observability matrices for each measurement

$$\mathbf{O}_{cur,k} = \frac{\alpha_x}{e_{cd,k}^2} \begin{bmatrix} y_k & -x_k & x_k^2 + y_k^2 \\ y_k & -x_k & \Sigma_k + x_k^2 + y_k^2 \\ y_k & -x_k & 2\Sigma_k + x_k^2 + y_k^2 \end{bmatrix},$$

$$\mathbf{O}_{tar,k} = \frac{\alpha_x}{y_k^2} \begin{bmatrix} y_k & -x_k & 0 \\ y_k & -x_k & \Sigma_k \\ y_k & -x_k & 2\Sigma_k \end{bmatrix},$$

where $\Sigma_k = y_k \Delta_{y,k} + x_k \Delta_{x,k}$ and $\Delta_{x,k} = T_s(\omega_k \ell \cos\phi_k + \upsilon_k \sin\phi_k)$, $\Delta_{y,k} = T_s \omega_k \ell \sin\phi_k - T_s \upsilon_k \cos\phi_k$. It can be seen that the matrix $\mathbf{O}_k^e = \left[\mathbf{O}_{cur,k}^T \; \mathbf{O}_{tar,k}^T \right]^T$ is of rank 2 and the linear approximation is not observable at each instant time. Thus, a local observability analysis is not enough to conclude about this property. The linearization can be seen as a piece-wise constant system for each instant time k, and the theory developed in [57] can be used (see appendix A.3.3 for an introduction). It can be verified that the null space basis of the matrix $\mathbf{O}_k^e$ is any state $\mathbf{x}_k = \lambda \left[x_k \; y_k \; 0 \right]^T$, where $\lambda \in R$. This subset of the state space satisfies $\mathbf{F}_k \mathbf{x}_k = \mathbf{x}_k$ and then, the observability can be determined through the stripped observability matrix (SOM) defined in A.8.

In order to get a smaller stripped observability matrix, we use the local observability matrix obtained from the target epipole ($\mathbf{O}_{tar,k}$). This is enough to conclude about the observability of the system with both measurements but also to achieve observability with only one measurement. This local observability matrix for the next instant time is

$$\mathbf{O}_{tar,k+1} = \frac{\alpha_x}{y_{k+1}^2} \begin{bmatrix} y_{k+1} & -x_{k+1} & 0 \\ y_{k+1} & -x_{k+1} & \Sigma_{k+1} \\ y_{k+1} & -x_{k+1} & 2\Sigma_{k+1} \end{bmatrix}.$$

Given that any non-null factor does not affect the rank of a matrix, we omit the different multiplicative factors of each local observability matrix to write the stripped observability matrix that is obtained in two steps as

$$\mathbf{O}_{SOM,1}^{e} = \begin{bmatrix} y_k & -x_k & 0 \\ y_k & -x_k & y_k\Delta_{y,k} + x_k\Delta_{x,k} \\ y_k & -x_k & 2\left(y_k\Delta_{y,k} + x_k\Delta_{x,k}\right) \\ y_k - \Delta_{y,k} & -x_k + \Delta_{x,k} & 0 \\ y_k - \Delta_{y,k} & -x_k + \Delta_{x,k} & \left(y_k - \Delta_{y,k}\right)\Delta_{y,k+1} + \left(x_k - \Delta_{x,k}\right)\Delta_{x,k+1} \\ y_k - \Delta_{y,k} & -x_k + \Delta_{x,k} & 2\left(\left(y_k - \Delta_{y,k}\right)\Delta_{y,k+1} + \left(x_k - \Delta_{x,k}\right)\Delta_{x,k+1}\right) \end{bmatrix}. \quad (4.6)$$

This matrix can be reduced by Gaussian elimination to a 3×3 triangular matrix whose determinant is $-2x_k^2\Delta_{x,k}\Delta_{y,k} + 2x_k y_k \Delta_{x,k}^2 - 2x_k y_k \Delta_{y,k}^2 + 2y_k^2\Delta_{x,k}\Delta_{y,k}$. Thus, under the assumption of sampling time different than zero, this matrix is full rank and the linear approximation $(\mathbf{F}_k, \mathbf{G}_k, \mathbf{H}_k^e)$ is observable if and only if non-null velocities are applied at each instant time. Moreover, a rotational velocity different than zero is enough to achieve observability if and only if $\ell \neq 0$, which agrees with the comments after lema 4.1. This analysis states that observability is gained in two steps even if local observability for each k is not ensured. □

It is worth emphasizing that both previous lemmas are valid for any pair of images, for instance, observability is also ensured by using the epipoles that relate the first captured image (initial image) and the current one. This property is exploited in order to solve the problem of short baseline when the target is being reached. It is known that the epipolar geometry relating the current and the target images becomes ill-conditioned when the cameras are very close each other. In this case, the epipoles are unstable and they are not useful as measurements anymore.

In order to avoid this problem, the measurements are switched to some new epipoles when an intermediate location aligned to the target without lateral error is reached. Thus, we exploit one of the benefits of the Kalman filtering approach, which allows to change the set of measurements online accordingly. The new epipoles are computed from the initial and the current images, and the intermediate location can be reached by tracking an adequate reference as will be described later (section 4.3.3). After the intermediate goal is reached, only a rectilinear forward motion remains to reach the target. So, the pose in this second stage is estimated from the epipoles relating the initial and the current images, which behave adequately.

Another important aspect in the extended Kalman filter implementation is related to the initial values of the estimated state. For this purpose, the initial pose can be recovered by decomposing the essential matrix like in the localization stage of the approach [43], which is based on the 5-point algorithm [113].

4.2.2 Observability Analysis with the 1D Trifocal Tensor as Measurement

The elements of the 1D trifocal tensor have shown to be useful providing information of position and orientation of a camera-robot through static methods, mainly for localization [49], [59]. The use of the 1D trifocal tensor allows to avoid the problem of short baseline without the need of switching the measurements. Additionally, given that the computation of the 1D trifocal tensor only needs the center of projection of the images, the resulting estimation scheme is semicalibrated. Moreover, as the 1D trifocal tensor is a more stable geometric constraint, the estimation scheme results more robust to image noise. This section analyzes the feasibility to implement an estimation scheme using one element of the 1D trifocal tensor as measurement with appropriate theoretical tools. Looking the expressions in (3.2), the elements of the tensor can be expressed related to the current location $\mathbf{C}_2 = (x, y, \phi)$ using a generic measurement model of the form

$$h^t(\mathbf{x}) = \alpha x \sin\phi + \beta x \cos\phi + \gamma y \sin\phi + \delta y \cos\phi, \tag{4.7}$$

where α, β, γ, δ are suitable constants defined for each tensor element and the superscript t refers to measurement from the tensor. This expression of the measurement model allows us to generalize the results for any tensor element.

4.2.2.1 Nonlinear Observability Analysis

Similarly to the analysis for the case of measurements from the epipoles, this section utilizes the theory of [65] (introduced in appendix A.3.1) to carry out the observability analysis. Since the analysis is more complex than for the epipoles because there are eight tensor elements, the following proposition is firstly established.

Proposition 4.1. *The space spanned by all possible Lie derivatives given by the generic measurement model (4.7) along the vector fields g_1 and g_2 of the continuous system (1.3) is of dimension three if the measurement accomplishes $\alpha + \delta \neq 0$ or $\beta - \gamma \neq 0$.*

Proof. This proposition is proved by finding the space spanned by all possible Lie derivatives and verifying its dimension. This space is given as

$$\Omega^t = \left(h^t, L_{g_1}^1 h^t, L_{g_2}^1 h^t, L_{g_1}^2 h^t, L_{g_2}^2 h^t, L_{g_1}L_{g_2}h^t, L_{g_2}L_{g_1}h^t, \ldots\right)^T. \tag{4.8}$$

The first order Lie derivatives are

$$\begin{aligned} L_{g_1}^1 h^t &= \delta \cos^2\phi - \alpha \sin^2\phi + (\gamma - \beta)\sin\phi\cos\phi = \varphi_a(\phi), \\ L_{g_2}^1 h^t &= -\ell\left(\beta\cos^2\phi + \gamma\sin^2\phi + (\alpha+\delta)\sin\phi\cos\phi\right) + \tfrac{\partial h^t}{\partial \phi}. \end{aligned}$$

We have introduced the notation φ for functions depending on ϕ, which emphasizes that some of the Lie derivatives only span in that direction. As a good approach, the search of functions in the Lie group is constrained for $n-1$, where $n = 3$ is the dimension of the state space. Then, the required second order Lie derivatives turn out to be

$$\begin{aligned} L_{g_1}^2 h^t &= 0, \\ L_{g_2}^2 h^t &= -\ell\left((2\alpha+\delta)\cos^2\phi + 3(\gamma-\beta)\sin\phi\cos\phi\right) - \ell(2\delta+\alpha)\sin^2\phi - h, \\ L_{g_1}L_{g_2}h^t &= \varphi_b(\phi), \\ L_{g_2}L_{g_1}h^t &= \varphi_c(\phi). \end{aligned}$$

In order to know the dimension of the space Ω, the gradient operator is applied to each one of the functions defining such space. Notice that the Lie derivatives $L_{g_1}L_{g_2}h$ and $L_{g_2}L_{g_1}h$ span in the same direction of $L_{g_1}^1 h$ and the corresponding gradients of the formers do not contribute to the dimension of the space. The dimension of the observable space is determined by the rank of the matrix

$$\nabla\Omega^t = \left[(\nabla h^t)^T \; (\nabla L_{g_1}^1 h^t)^T \; (\nabla L_{g_2}^1 h^t)^T \; (\nabla L_{g_2}^2 h^t)^T \right]^T \tag{4.9}$$

$$= \begin{bmatrix} \alpha s\phi + \beta c\phi & \gamma s\phi + \delta c\phi & \alpha x c\phi - \beta x s\phi + \gamma y c\phi - \delta y s\phi \\ 0 & 0 & -2(\alpha+\delta)s\phi c\phi + (\beta-\gamma)\left(s^2\phi - c^2\phi\right) \\ \alpha c\phi - \beta s\phi & \gamma c\phi - \delta s\phi & \ell\left((\alpha+\delta)\left(s^2\phi - c^2\phi\right) - 2(\beta-\gamma)s\phi c\phi\right) - h^t \\ -\alpha s\phi - \beta c\phi & -\gamma s\phi - \delta c\phi & \ell\left(6(\alpha+\delta)s\phi c\phi - 3(\beta-\gamma)\left(s^2\phi - c^2\phi\right)\right) - \frac{\partial h^t}{\partial\phi} \end{bmatrix}.$$

By taking, for instance, the first three rows, this matrix has rank three if $\alpha + \delta \neq 0$ or $\beta - \gamma \neq 0$. Thus, the space spanned by all possible Lie derivatives is of dimension three under such conditions. □

The interest of this analysis is to conclude about the observability property of the camera-robot pose using one element of the 1D trifocal tensor. In this sense, the following lema is stated.

Lemma 4.3. *The continuous camera-robot system (1.3) does not satisfy the observability rank condition of [65] (A.6) for any element of the 1D trifocal tensor (3.2) as measurement. Thus, locally weak observability cannot be ensured using information of the 1D trifocal tensor.*

Proof. This proof results as a derivation of the proposition 4.1. From (3.2), it is seen that, by including an additional term $\kappa f_{sc}(\phi)$ with κ a generic constant and $f_{sc}(\phi)$ being $s\phi$ or $c\phi$, the non-normalized elements of the tensor can be expressed in two of the following forms:

1. Any of the elements $T_{121}, T_{122}, T_{221}$ and T_{222} can be written as $h_1^t(x) = \beta x \cos\phi + \gamma y \sin\phi + \kappa f_{sc}(\phi)$. In accordance to the generic measurement model (4.7), for these tensor elements $\alpha = 0$, $\delta = 0$, $\beta = \gamma$, and consequently the conditions of the previous proposition are not accomplished.
2. The elements T_{111}, T_{112}, T_{211} and T_{212} can be expressed as $h_2^t(x) = \alpha x \sin\phi + \delta y \cos\phi + \kappa f_{sc}(\phi)$. In this case, $\beta = 0$, $\gamma = 0$, $\alpha = -\delta$, and the conditions in proposition 4.1 to span a space of dimension three are not fulfilled.

Hence, since $\alpha + \delta = 0$ and $\beta - \gamma = 0$ in any case, the observability matrix has only two rows linearly independent

$$\mathbf{O}_c = \begin{bmatrix} \nabla h^t \\ \nabla L_{g_2}^1 h^t \end{bmatrix} = \begin{bmatrix} \alpha \sin\phi + \beta \cos\phi & \gamma \sin\phi + \delta \cos\phi & \frac{\partial h^t}{\partial \phi} \\ \alpha \cos\phi - \beta \sin\phi & \gamma \cos\phi - \delta \sin\phi & -h^t \end{bmatrix}. \tag{4.10}$$

Given that this matrix $\mathbf{O}_c$ has a lower rank than the dimension of the state space ($n = 3$), the observability rank condition is not satisfied and consequently locally weak observability cannot be ensured for the continuous system (1.3) with any element of the 1D trifocal tensor (3.2) as measurement. The same result is obtained by using any linear combination of elements of the tensor. Higher order Lie derivatives do not modify this result because they are linearly dependent on lower order derivatives. In any case, the maximum rank of the observability matrix is two. □

Notice that in the observability matrix (4.10) only appears the gradient of the measurement along g_2, in addition to the gradient of h^t. This means that, since the vector field g_2 is associated to the rotational velocity, this velocity is important to provide a second observable direction in the state space. However, this is not enough to cover the three dimensional space. In contrast, the translational velocity does not contribute to gain any observable direction, because the gradient of any Lie derivative related to g_1 provides a linearly dependent row vector.

In spite of the previous continuous analysis is not able to state the complete observability of the camera-robot pose, the theory about discrete nonlinear observability developed in [112] and introduced in appendix A.3.2, allows us to enunciate the following result.

Lemma 4.4. *The discrete camera-robot system (1.5) is said to be observable according to definition A.7, [112], by using only one element of the 1D trifocal tensor (3.2) as measurement if rotational velocity is applied during two consecutive instant times and the corresponding velocities are different and no-null for these two consecutive steps.*

Proof. This is proved by constructing the corresponding nonlinear observability matrix and verifying its rank. Let us rewrite the generic measurement (4.7) to represent any of the eight elements of the 1D trifocal tensor in discrete time.

$$h^t(\mathbf{x}_k) = -\kappa_1 t_{x_k} + \kappa_2 t_{y_k} + \kappa_3 f_{sc}(\phi_k), \tag{4.11}$$

where $t_{x_k} = -x_k \cos\phi_k - y_k \sin\phi_k$, $t_{y_k} = x_k \sin\phi_k - y_k \cos\phi_k$ and related to (4.7) we have $\alpha = \kappa_2$, $\beta = \kappa_1$, $\gamma = \kappa_1$, $\delta = -\kappa_2$. In this case that the 1D trifocal tensor is used as measurement, the corresponding measurement matrix is given by:

$$\begin{aligned}\frac{\partial h^t}{\partial \mathbf{x}_k} &= \left[\, \kappa_1 \cos\phi_k + \kappa_2 \sin\phi_k \;\; \kappa_1 \sin\phi_k - \kappa_2 \cos\phi_k \;\; -\kappa_1 t_{y_k} - \kappa_2 t_{x_k} + \kappa_3 f_{cs}(\phi_k) \,\right] \\ &= \mathbf{H}_k^t = \left[\mathbf{H}_{x,k}^t \; \mathbf{H}_{y,k}^t \; \mathbf{H}_{\phi,k}^t \right]. \end{aligned} \tag{4.12}$$

The recursive operations described in definition A.7 result in the following nonlinear observability matrix, in which $\varepsilon_k = \phi_k + T_s\omega_k$ and $\zeta_k = \phi_k + T_s(\omega_k + \omega_{k+1})$:

$$\mathbf{O}_d = \begin{bmatrix} \kappa_1 c\phi_k + \kappa_2 s\phi_k & \kappa_1 s\phi_k - \kappa_2 c\phi_k & \kappa_1(-x_k s\phi_k + y_k c\phi_k) + \kappa_2(x_k c\phi_k + y_k s\phi_k) + \kappa_3 f_{cs}(\phi_k) \\ \kappa_1 c\varepsilon_k + \kappa_2 s\varepsilon_k & \kappa_1 s\varepsilon_k - \kappa_2 c\varepsilon_k & \kappa_1(-x_k s\varepsilon_k + y_k c\varepsilon_k) + \kappa_2(x_k c\varepsilon_k + y_k s\varepsilon_k) + \kappa_3 f_{cs}(\varepsilon_k) \\ \kappa_1 c\zeta_k + \kappa_2 s\zeta_k & \kappa_1 s\zeta_k - \kappa_2 c\zeta_k & \kappa_1(-x_k s\zeta_k + y_k c\zeta_k) + \kappa_2(x_k c\zeta_k + y_k s\zeta_k) + \kappa_3 f_{cs}(\zeta_k) \end{bmatrix}. \tag{4.13}$$

It can be seen that this matrix has three vector rows linearly independent if the following conditions are fulfilled:

$$T_s \neq 0,\ \omega_k \neq 0,\ \omega_{k+1} \neq 0,\ \omega_k \neq \omega_{k+1}. \tag{4.14}$$

Thus, it is proved that the observability matrix is full rank three and the system (1.5) is observable if rotational velocity is applied during two consecutive instant times and the corresponding velocities are different and no-null at each time. □

Notice that the difference between both previous observability analysis is that the continuous case considers only the information in a specific state of the system while in the discrete case, the analysis provides a way to introduce information from consecutive instant times. According to lemma 4.4, a digital implementation of an estimation scheme for the system (1.5) with measurement of the type (4.11) *collects* enough information along two instant times. In this sense, both results are complementary each other. The no locally weak observability states that the robot pose cannot be distinguished instantaneously; however, the pose can be estimated in two steps in accordance to the discrete analysis. The conditions for the observability in lemma 4.4 confirm the dependence of this property on the control inputs, in particular on the rotational velocity. Both continuous and discrete-time analysis agree each other with the fact that the translational velocity does not contribute to gain any observable direction, while the rotational velocity does.

4.2.2.2 Analysis as Piece-Wise Constant System

Up to now, we have proved that the camera-robot model with one element of the 1D trifocal tensor as measurement is an observable system from a nonlinear point of view. However, similarly to the case of epipolar measurements, the implementation is carried out through a linearization-based scheme as the EKF. In order to ensure a consistent estimation, it is important to verify the effect of the linearization over the observability of the state. One more time, we use the basic form of the EKF (see appendix A.4). The linear approximation $(\mathbf{F}_k, \mathbf{G}_k, \mathbf{H}_k^t)$ is treated as a piece-wise constant system so that the theory introduced in [57] is used.

Let us verify the condition that allows to test the observability from the stripped observability matrix (A.8). The local observability matrix for the k-th instant time is

$$\mathbf{O}_k = \begin{bmatrix} \mathbf{H}_{x,k}^t & \mathbf{H}_{y,k}^t & \mathbf{H}_{\phi,k}^t \\ \mathbf{H}_{x,k}^t & \mathbf{H}_{y,k}^t & \Lambda_k + \mathbf{H}_{\phi,k}^t \\ \mathbf{H}_{x,k}^t & \mathbf{H}_{y,k}^t & 2\Lambda_k + \mathbf{H}_{\phi,k}^t \end{bmatrix},$$

with $\Lambda_k = T_s\left(\kappa_2 \omega_k \ell - \kappa_1 \upsilon_k\right)$ and $\mathbf{H}_{x,k}$, $\mathbf{H}_{y,k}$, $\mathbf{H}_{\phi,k}$ as defined in (4.12). This is a matrix of rank two and its null space $NULL\left(\mathbf{O}_k\right)$ is any state $\mathbf{x}_k = \lambda \left[-\mathbf{H}_{y,k}\ \mathbf{H}_{x,k}\ 0\right]^T$, where $\lambda \in \mathbb{R}$. This subset of the state space satisfies $\mathbf{F}_k \mathbf{x}_k = \mathbf{x}_k$ and then, the observability can be determined through the stripped observability matrix (SOM). The local observability matrix for the next instant time is

$$\mathbf{O}_{k+1} = \begin{bmatrix} \mathbf{H}_{x,k+1}^t & \mathbf{H}_{y,k+1}^t & \mathbf{H}_{\phi,k+1}^t \\ \mathbf{H}_{x,k+1}^t & \mathbf{H}_{y,k+1}^t & \Lambda_{k+1} + \mathbf{H}_{\phi,k+1}^t \\ \mathbf{H}_{x,k+1}^t & \mathbf{H}_{y,k+1}^t & 2\Lambda_{k+1} + \mathbf{H}_{\phi,k+1}^t \end{bmatrix}.$$

Thus, the following stripped observability matrix $\mathbf{O}_{SOM,1}^t$, with $\varepsilon_k = \phi_k + T_s \omega_k$, is obtained in two steps:

$$\mathbf{O}_{SOM,1}^t = \begin{bmatrix} \kappa_1 c\phi_k + \kappa_2 s\phi_k & \kappa_1 s\phi_k - \kappa_2 c\phi_k & \mathbf{H}_{\phi,k}^t \\ \kappa_1 c\phi_k + \kappa_2 s\phi_k & \kappa_1 s\phi_k - \kappa_2 c\phi_k & T_s\left(\kappa_2 \omega_k \ell - \kappa_1 \upsilon_k\right) + \mathbf{H}_{\phi,k}^t \\ \kappa_1 c\phi_k + \kappa_2 s\phi_k & \kappa_1 s\phi_k - \kappa_2 c\phi_k & 2T_s\left(\kappa_2 \omega_k \ell - \kappa_1 \upsilon_k\right) + \mathbf{H}_{\phi,k}^t \\ \kappa_1 c\varepsilon_k + \kappa_2 s\varepsilon_k & \kappa_1 s\varepsilon_k - \kappa_2 c\varepsilon_k & \mathbf{H}_{\phi,k+1}^t \\ \kappa_1 c\varepsilon_k + \kappa_2 s\varepsilon_k & \kappa_1 s\varepsilon_k - \kappa_2 c\varepsilon_k & T_s\left(\kappa_2 \omega_{k+1} \ell - \kappa_1 \upsilon_{k+1}\right) + \mathbf{H}_{\phi,k+1}^t \\ \kappa_1 c\varepsilon_k + \kappa_2 s\varepsilon_k & \kappa_1 s\varepsilon_k - \kappa_2 c\varepsilon_k & 2T_s\left(\kappa_2 \omega_{k+1} \ell - \kappa_1 \upsilon_{k+1}\right) + \mathbf{H}_{\phi,k+1}^t \end{bmatrix}. \tag{4.15}$$

By looking at the conditions that make the matrix $\mathbf{O}_{SOM,1}^t$ full rank and comparing with those conditions stated in lemma 4.4, the following corollary can be enunciated.

Corollary 4.1. *The piece-wise constant system given by the linear approximation* $(\mathbf{F}_k, \mathbf{G}_k, \mathbf{H}_k^t)$ *of the system (1.5) and measurement from the 1D trifocal tensor (4.12), as used by the EKF, is observable under the same conditions stated in lemma 4.4. Under such conditions, the matrix* $\mathbf{O}_{SOM,1}^t$ *(4.15) is full rank.*

It is worth mentioning that the results on observability of this section are valid for any normalized element of the tensor, except for T_{121}, which is equal to one after the normalization. Although the previous mathematical development has been shown for the non-normalized tensor for clarity, the implementation of the EKF considers normalized measurements.

4.2.2.3 Initialization of the Estimation

Additional to the initial values of the estimated state, in our framework, the measurement Jacobian (4.12) requires to know the initial location $\mathbf{C}_1$. We describe a method that obtains all the required initial information through the 1D trifocal tensor. It is known that estimating the 1D trifocal tensor from a set of images where two of them are exactly the same is numerically troublesome, which is the case for the initial condition. In order to avoid this problem and also obtain a useful metric information, a third image is captured after an initial forward motion. Thus, the complete geometry of the three views is estimated by knowing the relative location of one of the images with respect to another. Because of the initial motion, we know the relative position between $\mathbf{C}_1$ and $\mathbf{C}_2$ and also that $\phi_1 = \phi_2 = \phi$. The following system of equations gives the position up to scale and orientation of $\mathbf{C}_3$ with respect to $\mathbf{C}_1$:

$$\begin{bmatrix} x \\ y \\ \alpha s\phi \\ \alpha c\phi \end{bmatrix} = \frac{1}{a_1^2+a_2^2} \begin{bmatrix} \frac{a_1 a_2}{a_4} & \frac{a_1 a_2}{a_4} & \frac{a_1^2}{a_4} & -\frac{a_2^2}{a_4} \\ \frac{a_1^2}{a_4} & \frac{a_2^2}{a_4} & \frac{a_1 a_2}{a_4} & \frac{a_1 a_2}{a_4} \\ a_1 & a_1 & -a_2 & -a_2 \\ a_2 & a_2 & a_1 & a_1 \end{bmatrix} \begin{bmatrix} T_{112} \\ T_{121} \\ T_{212} \\ T_{221} \end{bmatrix}. \tag{4.16}$$

The choice of the tensor elements in (4.16) is the unique possible in order to have a non-singular system of equations, given that $a_1 = 0$, $a_2 = d_{ini}$, $a_3 = 0$, $a_4 = 1$, with d_{ini} being the longitudinal distance in the initial motion. These values are obtained by analyzing the particular configuration of the locations with reference in $\mathbf{C}_1$ to facilitate the derivations. Once the orientation ϕ is known, the scale factor (σ) can be estimated from $\sigma T_{211} = -t_{x_1} \sin\phi_2 - t_{y_2} \cos\phi_1$, where $\phi_1 = -\phi$, $\phi_2 = 0$, $t_{y_2} = d_{ini}$ and the value of t_{x_1} is not important. Thus, the estimated scale factor is

$$\sigma = -\frac{d_{ini} \cos\phi}{T_{211}}.$$

Recall that finally the coordinates of $\mathbf{C}_1$ and $\mathbf{C}_2$ must be expressed with respect to $\mathbf{C}_3$. Now, the required information, $\mathbf{C}_1$ for the measurement Jacobian (4.12) and $\mathbf{C}_2$ for the EKF initialization, has been deduced.

4.3 Non-holonomic Visual Servoing in the Cartesian Space

Similarly to previous chapters, the goal in this section is to drive a mobile robot to a target location, i.e., to reach a desired position and orientation $\mathbf{C}_2 = (0,0,0)$. In spite of the Brockett's theorem is not satisfied by non-holonomic mobile robots, in this section we solve the pose regulation problem for these robots by using a smooth estimated-state feedback control law and adequate references to track. The controller drives the lateral and longitudinal robot positions to zero by tracking desired references. At the same time, orientation correction is also achieved through smooth input velocities.

The location of the camera mounted on the robot is controlled instead of the robot reference frame because the measurements are given with respect to the camera frame. The controller is designed from the discrete model (1.4) considering that the feedback information is given by the estimated state $\hat{\mathbf{x}}_k = \left[\hat{x}_k \ \hat{y}_k \ \hat{\phi}_k\right]^T$ obtained as described in lemma 4.2 or corollary 4.1. The important aspect in control theory regarding to the stability of the closed loop system using feedback of the estimated state is addressed in this section. In this sense, the validity of a separation principle for our nonlinear framework is verified.

The advantage of the presented control scheme with respect to all previous works on visual servoing for mobile robots is that the real-world path followed by the platform can be predefined. This is possible because the state estimation allows us to tackle the visual servoing problem as a trajectory tracking in the Cartesian space. It provides the advantage to facilitate the planning of complex tasks, like homing for large displacements, path following or reactive navigation [52].

4.3.1 Control of the Position Error

This section presents a solution to the problem of output feedback tracking using the input-output linearization control technique (see appendix A.1 and [124] for details). Because of a mobile robot is a underactuated system, by controlling the lateral and longitudinal position's coordinates, the orientation (ϕ) remains as a degree of freedom of the control system. Nevertheless, the orientation can be simultaneously corrected by tracking suitable desired trajectories. Thus, a discrete linearizing controller to take the value of the robot position to zero in a smooth way is described. Let us define the output to be controlled as the reduced state vector

$$\mathbf{x}_{r,k} = \left[x_k \ y_k\right]^T. \tag{4.17}$$

Hence, the tracking errors are $\xi_k^1 = x_k - x_k^d$, $\xi_k^2 = y_k - y_k^d$, where x_k^d and y_k^d are the discrete values of the desired smooth time-varying references that are defined later. The difference equations of these errors result in the system

$$\begin{bmatrix} \xi_{k+1}^1 \\ \xi_{k+1}^2 \end{bmatrix} = \begin{bmatrix} \xi_k^1 \\ \xi_k^2 \end{bmatrix} + T_s \begin{bmatrix} -\sin\phi_k & -\ell\cos\phi_k \\ \cos\phi_k & -\ell\sin\phi_k \end{bmatrix} \begin{bmatrix} v_k \\ \omega_k \end{bmatrix} - T_s \begin{bmatrix} \dot{x}_k^d \\ \dot{y}_k^d \end{bmatrix}. \tag{4.18}$$

This position error system can be expressed as

$$\xi_{k+1} = \xi_k + T_s \mathbf{D}(\phi_k, \ell)\,\mathbf{u}_k - T_s \dot{\mathbf{x}}_{r,k}^d, \tag{4.19}$$

where $\xi_k = \mathbf{x}_{r,k} - \mathbf{x}_{r,k}^d$, $\mathbf{D}(\phi_k, \ell)$ is the decoupling matrix depending on the orientation and the fixed parameter ℓ and $\dot{\mathbf{x}}_{r,k}^d = \left[\dot{x}_k^d, \dot{y}_k^d\right]^T$ represents a known perturbation for the error dynamics. The corresponding inverse matrix to decouple the system is given as

$$\mathbf{D}^{-1}(\phi_k, \ell) = \frac{1}{\ell} \begin{bmatrix} -\ell\sin\phi_k & \ell\cos\phi_k \\ -\cos\phi_k & -\sin\phi_k \end{bmatrix}. \tag{4.20}$$

Given that the control inputs appear in the first differentiation of each output, the camera-robot system (1.4) with position coordinates as outputs (4.17) has a vector relative degree $\{1,1\}$. Then, the sum of the indices of the system $(1+1)$ is less than the dimension of the state space $(n=3)$ and a first order zero dynamics appears, which represents the previously mentioned degree of freedom (ϕ) of the system.

As the control is based on estimation, the static state feedback control law $\mathbf{u}_k$ resulting from the inversion of the error system (4.19) turns out to be

$$\hat{\mathbf{u}}_k = \begin{bmatrix} \hat{\upsilon}_k \\ \hat{\omega}_k \end{bmatrix} = \mathbf{D}^{-1}\left(\hat{\phi}_k, \ell\right) \begin{bmatrix} \hat{v}_k^1 \\ \hat{v}_k^2 \end{bmatrix}, \tag{4.21}$$

where $\hat{v}_k^1 = -\lambda_1 \hat{\xi}_k^1 + \dot{x}_k^d$ and $\hat{v}_k^2 = -\lambda_2 \hat{\xi}_k^2 + \dot{y}_k^d$. It can be verified that the input velocities achieve global stabilization of the position error system (4.19) in the case of feedback of the real state $\mathbf{u}_k$. In such a case, the dynamic behavior of the closed loop position error is exponentially stable if and only if λ_1 and $\lambda_2 \in (0,\ 2/T_s)$. In the subsequent section, the stability of the closed loop system with feedback of the estimated state $\hat{\mathbf{u}}_k$ is analyzed.

Note that this input-output linearization via static feedback is only possible for the camera-robot system (1.4) with known $\ell \neq 0$. Otherwise, a singular decoupling matrix is obtained and a static feedback fails to solve the input-output linearization problem. Nevertheless, the case of having the camera shifted from the robot rotational axis over the longitudinal axis is a common situation.

4.3.2 Stability of the Estimation-Based Control Loop

In this section, the stability of the closed loop with feedback of the estimated camera-robot pose is studied. It is well known that the separation principle between estimation and control does not apply for nonlinear systems, however, it has been investigated for a class of nonlinear systems [8], which can be expressed in a nonlinear canonical form. As the position error dynamics (4.19) does not lie in that class of systems, then we present a particular analysis.

Let us first obtain the dynamic system of the estimation error and subsequently study the interaction of this error with the closed loop control system. The dynamic behavior of the a priori estimation error is given by

$$\mathbf{e}_{k+1}^- = \mathbf{x}_{k+1} - \hat{\mathbf{x}}_{k+1}^- = f\left(\mathbf{x}_k, \hat{\mathbf{u}}_k\right) - f\left(\hat{\mathbf{x}}_k^+, \hat{\mathbf{u}}_k\right). \tag{4.22}$$

It is worth emphasizing that the same conclusions of the analysis can be obtained using the a posteriori estimation error $\mathbf{e}_k^+$ [117]. Let us introduce some expressions to expand the smooth nonlinear functions f and h, with the last being a generic function representing (4.2) or (4.11)

$$f\left(\mathbf{x}_k, \hat{\mathbf{u}}_k\right) - f\left(\hat{\mathbf{x}}_k^+, \hat{\mathbf{u}}_k\right) = \mathbf{F}_k\left(\mathbf{x}_k - \hat{\mathbf{x}}_k^+\right) + \Phi\left(\mathbf{x}_k, \hat{\mathbf{x}}_k^+, \hat{\mathbf{u}}_k\right), \tag{4.23}$$

$$h\left(\mathbf{x}_k\right) - h\left(\hat{\mathbf{x}}_k^-\right) = \mathbf{H}_k\left(\mathbf{x}_k - \hat{\mathbf{x}}_k^-\right) + \Psi\left(\mathbf{x}_k, \hat{\mathbf{x}}_k^-\right), \tag{4.24}$$

with matrices $\mathbf{F}_k$ and $\mathbf{H}_k$ the Jacobians of the corresponding functions. By substituting (4.23) into (4.22) and using the discrete observer as given by the update stage of the EKF

$$\hat{\mathbf{x}}_k^+ = \hat{\mathbf{x}}_k^- + \mathbf{K}_k \left(h\left(\mathbf{x}_k\right) - h\left(\hat{\mathbf{x}}_k^-\right)\right),$$

we have

$$\mathbf{e}_{k+1}^- = \mathbf{F}_k \left(\mathbf{x}_k - \hat{\mathbf{x}}_k^- - \mathbf{K}_k \left(h\left(\mathbf{x}_k\right) - h\left(\hat{\mathbf{x}}_k^-\right)\right)\right) + \Phi\left(\mathbf{x}_k, \hat{\mathbf{x}}_k^+, \hat{\mathbf{u}}_k\right).$$

By substituting (4.24) and knowing that the a priori estimation error is given as $\mathbf{e}_k^- = \mathbf{x}_k - \hat{\mathbf{x}}_k^-$, then

$$\mathbf{e}_{k+1}^- = \mathbf{F}_k \left(\mathbf{I}_3 - \mathbf{K}_k \mathbf{H}_k\right) \mathbf{e}_k^- + \Theta_k,$$

where $\Theta_k = \Phi\left(\mathbf{x}_k, \hat{\mathbf{x}}_k^+, \hat{\mathbf{u}}_k\right) - \mathbf{F}_k \mathbf{K}_k \Psi\left(\mathbf{x}_k, \hat{\mathbf{x}}_k^-\right)$. Let us denote the first two components of the vector $\mathbf{e}_k^-$ as $\mathbf{e}_{r,k}^-$. The estimated tracking error ($\hat{\xi}_k$) is related to this reduced vector of estimation errors as follows:

$$\hat{\xi}_k = \xi_k - \mathbf{e}_{r,k}^-.$$

The control law (4.21), with $\hat{\mathbf{D}}^{-1} = \mathbf{D}^{-1}\left(\hat{\phi}_k, \ell\right)$ to simplify the notation, can be written using the estimated tracking error

$$\hat{\mathbf{u}}_k = \hat{\mathbf{D}}^{-1}\left(-\mathbf{k}\left(\xi_k - \mathbf{e}_{r,k}^-\right) + \dot{\mathbf{x}}_{r,k}^d\right).$$

By introducing this control input in the tracking error system (4.19), the closed loop difference equation with estimated state feedback results

$$\xi_{k+1} = \left(\mathbf{I}_2 - T_s \mathbf{D}\hat{\mathbf{D}}^{-1}\mathbf{k}\right)\xi_k + T_s \mathbf{D}\hat{\mathbf{D}}^{-1}\mathbf{k}\mathbf{e}_{r,k}^- + T_s\left(\mathbf{D}\hat{\mathbf{D}}^{-1} - \mathbf{I}_2\right)\dot{\mathbf{x}}_{r,k}^d.$$

The product of matrices

$$\mathbf{D}\hat{\mathbf{D}}^{-1} = \begin{bmatrix} \sin\phi_k \sin\hat{\phi}_k + \cos\phi_k \cos\hat{\phi}_k & -\sin\phi_k \cos\hat{\phi}_k + \sin\hat{\phi}_k \cos\phi_k \\ -\sin\hat{\phi}_k \cos\phi_k + \sin\phi_k \cos\hat{\phi}_k & \cos\phi_k \cos\hat{\phi}_k + \sin\phi_k \sin\hat{\phi}_k \end{bmatrix}$$

turns out to be a definite positive matrix with $\det(\mathbf{D}\hat{\mathbf{D}}^{-1}) = \left(\sin\phi_k \sin\hat{\phi}_k\right)^2 + \left(\cos\phi_k \cos\hat{\phi}_k\right)^2 + \left(\sin\phi_k \cos\hat{\phi}_k\right)^2 + \left(\sin\hat{\phi}_k \cos\phi_k\right)^2 > 0$ and $\mathbf{D}\hat{\mathbf{D}}^{-1}$ becomes the identity if $\phi_k = \hat{\phi}_k$. Finally, the overall closed loop control system with estimated state feedback is expressed as follows:

$$\begin{bmatrix} \xi_{k+1} \\ \mathbf{e}_{k+1}^- \end{bmatrix} = \begin{bmatrix} \mathbf{I}_2 - T_s \mathbf{D}\hat{\mathbf{D}}^{-1}\mathbf{k} & \left[T_s \mathbf{D}\hat{\mathbf{D}}^{-1}\mathbf{k}, 0\right] \\ \mathbf{0} & \mathbf{F}_k\left(\mathbf{I}_3 - \mathbf{K}_k \mathbf{H}_k\right) \end{bmatrix} \begin{bmatrix} \xi_k \\ \mathbf{e}_k^- \end{bmatrix} + \begin{bmatrix} T_s\left(\mathbf{D}\hat{\mathbf{D}}^{-1} - \mathbf{I}_2\right)\dot{\mathbf{x}}_{r,k}^d \\ \mathbf{r}_k \end{bmatrix}.$$

The triangular form of this system shows that the stability property can be achieved ensuring the stability of each one of the dynamics ξ_k and $\mathbf{e}_k^-$, i.e., a separation principle holds for the system. Notice that each dynamics is subject to perturbations. The tracking error is subject to a perturbation depending on the derivative of the desired references and the estimation error is subject to a vanishing perturbation depending on its own dynamics. The more important effect comes from the second one, because the former perturbation depends on the accuracy of the orientation estimation and even, it can be neglected considering the smoothness of the reference signals. Thus, the stability of the overall control scheme is determined by the estimation error dynamics. According to Theorem 7 stated in [117], the extended Kalman filter behaves as an exponential observer given the boundness of the matrices of the linear approximation, the boundness of the estimation covariances, the nonsingularity of the matrix $\mathbf{F}_k$ and boundness of the perturbation Θ_k. These conditions of convergence are accomplished in our system, and consequently, exponential stability of the overall control system is achieved.

It is worth noting that, as any dynamic observer, the EKF needs the value of the robot velocities $\mathbf{u}_k$. In our framework of visual control, the velocities given to the robot are known without the need of measuring them. Moreover, any uncertainty in $\mathbf{u}_k$ can be considered as noise in the state of the camera-robot system and assumed by the corresponding covariance.

4.3.3 Pose Regulation through Adequate Reference Tracking

The pose regulation problem requires to reach a desired position and orientation with respect to a fixed reference frame, which in our case is defined by the location associated to the target image. Up to now, the presented controller drives to zero the lateral and longitudinal errors through a smooth evolution, but the orientation evolves freely. In order to obtain also orientation correction for non-holonomic mobile robots, a good option is to define an adequate path for the robot position. The following time-differentiable references are used to achieve the desired behavior of the robot position and consequently to reach the desired orientation

$$
\begin{aligned}
y_k^d &= \frac{y^i - y^f}{2}\left(1+\cos\left(\frac{\pi}{\tau_{st}}kT_s\right)\right) + y^f, && 0 \le kT_s \le \tau_{st}, \\
x_k^d &= \frac{x^i - x^f}{(y^i - y^f)^2}\left(y_k^d - y^f\right)^2 + x^f, && 0 \le kT_s \le \tau_{st},
\end{aligned} \tag{4.25}
$$

where (x^i, y^i) is the initial position and (x^f, y^f) is the desired final position that is reached in τ_{st} seconds. The subscript *st* refers to stage, given that it may be needed several stages to eventually reach the target pose. For instance, when the epipoles are used it is needed to define an intermediate goal without lateral error before the target. Thus, minimum two stages are defined in such case. Similarly to how an intermediate aligned goal is defined, we are able to define any other intermediate location through the parabolic path. So, this provides the possibility to avoid an obstacle detected along the path toward the target.

Additionally, the tracking of these references allows to define a fixed temporal horizon τ (the sum of the time of each stage) to reach the target location. Notice that these references depict a parabolic path on the $x-y$ plane from the point (x^i, y^i), which corresponds to the estimated starting position $\mathbf{C}_2$ as obtained from the initialization procedure. Thus, the robot always begins over the desired path and the controller has to maintain it tracking that path. Because of the non-holonomic motion of the robot, the reference tracking drives the robot to perform an initial rotation autonomously in order to be aligned with the path.

As mentioned previously, when the controlled outputs reach zero at the time τ the so-called *zero dynamics* is achieved in the system. In the particular case of the camera-robot system (1.4) with outputs $s_1 = x_k$, $s_2 = y_k$, this set is given as

$$Z^* = \left\{ \begin{bmatrix} x_k & y_k & \phi_k \end{bmatrix}^T \mid s_1 \equiv 0,\ s_2 \equiv 0 \right\} = \left\{ \begin{bmatrix} 0 & 0 & \bar{\phi}_k \end{bmatrix}^T,\ \bar{\phi}_k = \text{constant} \in \mathbb{R} \right\}.$$

The constant value of the orientation $\bar{\phi}_k$ is the solution to the following difference equation that characterize the zero dynamics:

$$\phi_{k+1} - \phi_k = -\frac{1}{\ell}\left(\hat{v}_k^1 \cos\phi_k + \hat{v}_k^2 \sin\phi_k\right) = 0,$$

because $\hat{v}_k^1 = 0$ and $\hat{v}_k^2 = 0$ when $s_1 \equiv 0$ and $s_2 \equiv 0$. Thus, zero dynamics in this control system means that when lateral and longitudinal positions of the camera-robot system are corrected, the orientation may be different to zero. Next, it is proved that orientation correction is also achieved by tracking the described references, in such a way that pose regulation is achieved.

Proposition 4.2. *The control inputs (4.21) with feedback of the estimated state* $\hat{\mathbf{x}}_k = \begin{bmatrix} \hat{x}_k & \hat{y}_k & \hat{\phi}_k \end{bmatrix}^T$ *provided by the estimation scheme described in lemma 4.2 or corollary 4.1 and using the reference signals (4.25), drive the camera-robot system (1.4) to reach the location* ($x = 0$, $y = 0$, $\phi = 0$)*, i.e., orientation is also corrected.*

Proof. In the previous section we have proved the stability of the position error dynamics with feedback of the estimated state, in such a way that correction of the lateral and longitudinal errors is ensured in τ seconds. It only remains to prove that the orientation is also zero when the target location is reached. From the decomposition of the translational velocity vector given by the kinematic behavior of the robot and using the difference equations $x_{k+1} - x_k = -\delta\,\hat{\upsilon}_k \sin\phi_k$, $y_{k+1} - y_k = \delta\,\hat{\upsilon}_k \cos\phi_k$, we have that

$$\phi_k = \arctan\left(-\frac{x_{k+1} - x_k}{y_{k+1} - y_k}\right).$$

Let us define the parabolic relationship between Cartesian coordinates $x = \frac{x_0}{y_0^2} y^2$ according to the desired trajectories (4.25). Its corresponding discrete time-derivative results in $x_{k+1} - x_k = 2\frac{x_0}{y_0^2} y_k (y_{k+1} - y_k)$. Thus, when the x and y-coordinates track the desired trajectories, the robot orientation is related to the current lateral position as follows:

$$\phi_k = \arctan\left(-2\frac{x_0}{y_0^2} y_k\right).$$

As mentioned, when the robot has followed the reference path and $kT_s = \tau$ the position reaches zero ($x = 0$, $y = 0$), and consequently $\phi = \arctan(0) = 0$. This proves that although the orientation is a degree of freedom for the control system, the location $(x = 0, y = 0, \phi = 0)$ is reached in τ seconds by tracking the defined profile (4.25) for the position coordinates. □

This behavior can be obtained whenever the tangent of the path is zero at the origin, as in (4.25). Thus, it is possible to use different functions besides a parabolic one in order to ensure that the robot reaches the target with the desired orientation, for instance, $x^d = x_0\left(1 - \cos\left(y^d \pi / 2y_0\right)\right)$. However, a smoother initial performance of the robot motion is obtained using the parabolic path.

Note that pose regulation is achieved using a single controller and smooth control inputs. Additionally, the approach takes into account the nonholomicity of wheeled mobile robots unlike [100]. We claim that as well as solving the pose regulation problem, the presented visual servoing scheme can be extended, for instance, to navigation from a visual memory.

4.4 Experimental Evaluation

This section presents an evaluation of the complete approach previously detailed, first, through simulations for different types of central cameras. Then, its validity is verified in real-world experiments using our experimental platform, the robot Pioneer 3-AT from Adept Mobile Robots that is shown in Fig. 1.1(a), equipped with the hypercatadioptric imaging system of Fig. 1.2(b). In both, simulations and experiments, the performance of the estimation scheme and of the pose controller is presented. This evaluation shows that the desired pose is always reached.

4.4.1 Simulation Results

Simulations have been performed in Matlab, where the geometric constraints are estimated from virtual omnidirectional images. The distance from the rotation axis to the camera position on the robot is set to $\ell = 8$ cm. For the controllers, the control gains are set to $\lambda_1 = 1$, $\lambda_2 = 1$. The sampling time of the control loop T_s is set to 0.5 s. Related to the Kalman filtering, the matrices $\mathbf{M}_k$ and $\mathbf{N}_k$ have been fixed accordingly by using small standard deviations in $\mathbf{N}_k$ and similar standard deviations in $\mathbf{M}_k$. An image noise with standard deviation of 1 pixel has been added, which produces a significative noise in the measurements. Initial standard deviations for the state estimation errors have been suggested as $\mathbf{P}_0 = diag(5^2\ \text{cm}, 10^2\ \text{cm}, 1^2\ \text{deg})$, which roughly reflect the error in the estimation of the initial values given by the initialization procedure.

4.4.1.1 Performance of the Estimation Scheme

In this section, we show an example of the performance of the estimation scheme from measurements of the 1D trifocal tensor, given that the results are analogous using the epipoles. The results are for the initial location $(-8,-12,-20^o)$ although they are similar for any case. As an example of the measurements used in the estimation process, the evolution of the elements of the trifocal tensor is displayed in Fig. 4.1. The state estimation errors of Fig. 4.2(a) are obtained by taking the element T_{111} as measurement. The figure presents the average errors and the average uncertainty bounds over all 100 Monte Carlo runs for each time step. The instantaneous errors are computed as the difference between the truth state given by the camera-robot model and the estimated state. It can be seen that each one of the three estimation errors are maintained within the 2σ confidence bounds.

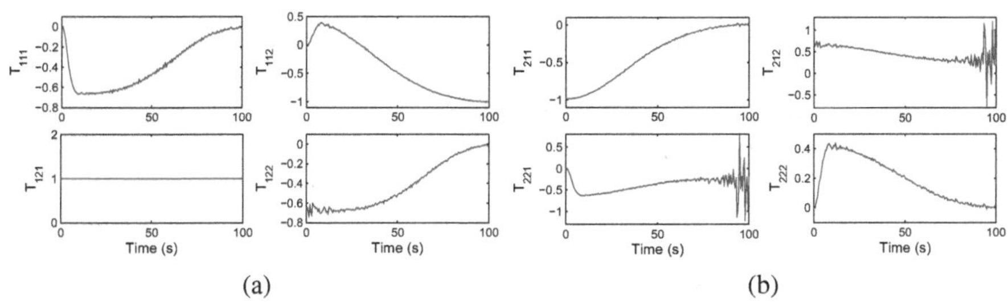

Fig. 4.1 Example of the evolution of the normalized elements of the tensor for the initial location (-8,-12,-20°). (a) First four tensor elements. (b) Second four tensor elements. T_{111} is used as measurement for the estimation. Notice that although any tensor element can be taken as measurement, T_{212} and T_{221} are particularly troublesome because they exhibit an unstable behavior at the end of the task.

We have also carried out a consistency test in order to determine if the computed covariances match the actual estimation errors. To do that, the consistency indexes normalized estimate error square (NEES, (A.11)) and normalized innovation squared (NIS, (A.12)) are used. Fig. 4.2(b) shows the average indexes NEES and NIS over the same 100 Monte Carlo runs as for the estimation errors. According to this figure, where the NEES and NIS indexes are less than one, the EKF with the chosen measurement is always consistent in spite of the nonlinearities of the state model and of the measurement model.

Robustness to vertical camera alignment and center of projection.

The estimation scheme that exploits the 1D trifocal tensor as measurements provides benefits of robustness against uncertainty in parameters, given that in the omnidirectional case the scheme using the epipolar geometry requires to know the calibration parameters and to ensure the vertical alignment of the camera. Fig. 4.3 presents the performance of the estimation scheme from the 1D trifocal tensor for variation of the vertical alignment and the center of projection. The figures depict the mean and standard deviation of the mean squared error for each state variable over 50 Monte Carlo runs. In order to discard any effect of the distribution of image features, the 3D scene in this simulation is a random distribution of points for each Monte Carlo run.

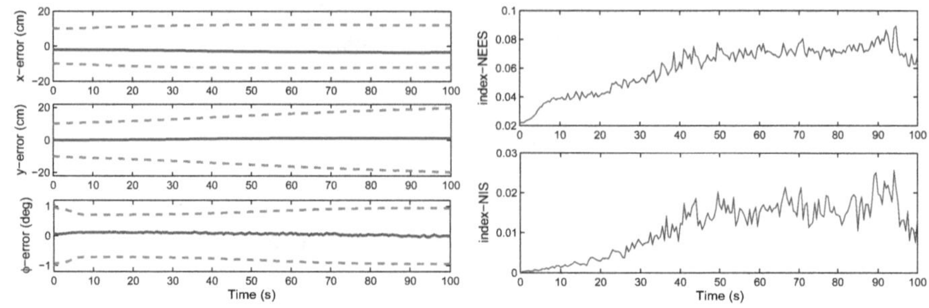

(a) (b)

Fig. 4.2 Example of performance of the estimation process obtained from Monte Carlo simulations with the 1D trifocal tensor as measurement. (a) State estimation errors and 2σ uncertainty bounds. (b) Consistency of estimation. Although this is for the initial location $(-8, -12, -20^o)$, similar results are obtained in any case.

It can be seen from both figures that the effect of disalignment of the camera and variation of the center of projection over the estimation errors is small, which provides good robustness of the estimation scheme against these aspects. This result is a consequence of the small effect of varying these parameters over the computation of the bearing measurements.

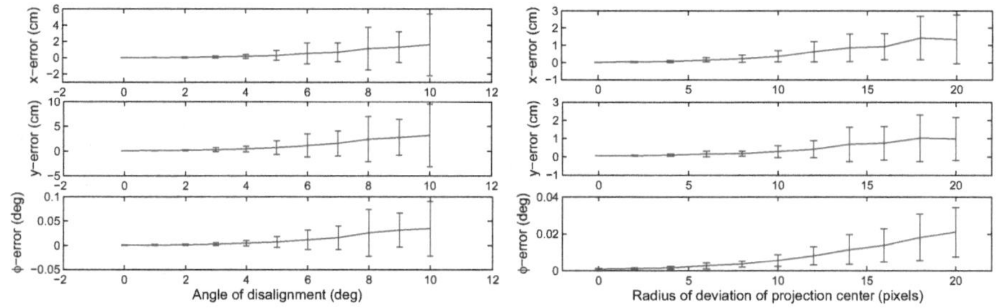

(a) (b)

Fig. 4.3 Robustness of the estimation against (a) disalignment of the camera and, (b) variation of the center of projection from Monte Carlo simulations with the 1D trifocal tensor as measurement

4.4.1.2 Closed Loop Performance Using the Estimated Pose

This section evaluates the validity of using the estimated pose in visual servoing tasks from different initial locations, firstly, using the epipolar geometry and secondly, exploiting the 1D trifocal tensor. The control law is used as established in the proposition 4.2. The synthetic images have been generated from the same 3D scene used for the simulations in chapter 3 (Fig. 2.4).

Estimation from the epipoles.

The size of the images used is 800×600 and the time τ to complete the whole task is fixed to 120 s for three different initial locations (8,-8,0^o), (2,-12,45^o) and (-6,-16,20^o). Fig. 4.4(a) shows an upper view of the robot motion on the plane for each initial location. It can be seen that in each case an initial rotation is carried out autonomously to align the robot with the parabolic path (4.25) to be tracked. In the three cases the robot is successfully driven to the target and for the last case, a fixed obstacle is also avoided by defining accordingly an intermediate goal position using the reference path (4.25). We assume that the obstacle detection is provided in time in order to modify the reference path as required. In Fig. 4.4(b), it can be seen that the position and orientation reach to zero at the end, but firstly, the longitudinal position (y) reaches -2 m at 100 s. After that, the measurements are changed to avoid the short baseline problem.

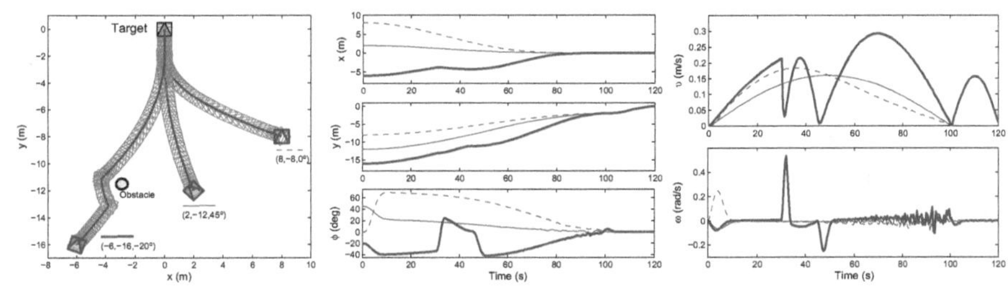

(a) (b) (c)

Fig. 4.4 Simulation results for some visual servoing tasks. The motion of the robot starts from three different initial locations and for one of them, an obstacle avoidance is carried out. (a) Paths on the $x-y$ plane. (b) State variables of the robot. (c) Computed velocities.

The two stages of the control task can be appreciated in the computed velocities shown in Fig. 4.4(c). From 0 s to 100 s, adequate velocities are computed for each case and particularly, it is worth commenting the case with obstacle avoidance. During the first seconds, the robot aligns with the initial reference path by rotating and follows that path until the obstacle is detected at 32 s, which interrupts the task. In this moment, the robot stops and rotates to the left and then, it follows a new path until that a desired intermediate goal is reached at 45 s. In that point, the robot stops and rotates to the right to start following a new path until 100 s, when it achieves to aligned to the target without lateral error. Finally, from rest, the robot moves forward for 20 s to correct the remaining longitudinal error by tracking the same sinusoidal profile for each case. Thus, the same translational velocity is applied for the final rectilinear motion in the three cases. It is worth noting that the velocities excite to the system adequately, in such a way that observability is always ensured.

The results of the previous figures have been obtained for hypercatadioptric and parabolic imaging systems. As an example, Fig. 4.5(a) shows the motion of the image points for the case (-6,-16,-20°) for a hypercatadioptric system. The corresponding epipoles as computed from twelve image points along the sequence are shown in Fig. 4.5(b)-(c) for each initial location. The epipole e_{tar} is the one used as measurement for the estimation during the first 100 s and after that, when it becomes unstable, the epipole e_{ini} is used for the last 20 s. Notice that even that no rotation is carried out during these last seconds, the measurement e_{ini} changes as the robot moves and then, observability is achieved given that the translational velocity is non-null.

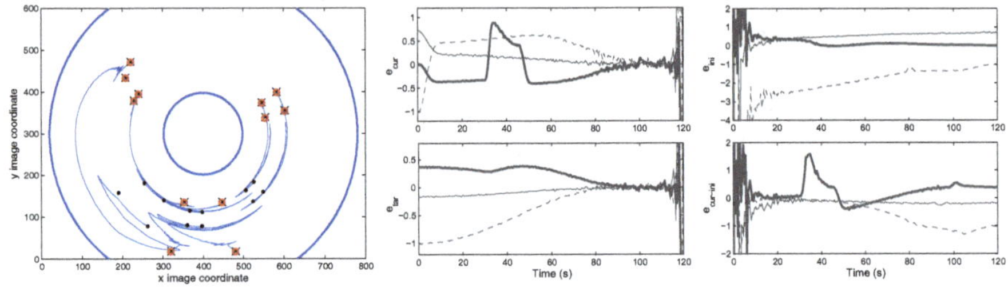

(a) (b) (c)

Fig. 4.5 Example of the behavior of the control scheme. (a) Motion of the image points. The black points represent the initial image, the red circles represent the target image and the times signs represent the image at the end of the motion. (b) Epipoles current-target. (c) Epipoles initial-current.

Estimation from the 1D trifocal tensor.

As proven theoretically in section 4.2.2, any tensor element can be used as measurement and the same observability properties are obtained. However, T_{111} is chosen for the evaluation since it has shown a good behavior during a servoing task in the previous chapter. In this case, the size of the images is 1024$\times$768 and the time τ to complete the whole task is fixed to 100 s. Fig. 4.6(a) shows the paths traced along the motion of the robot from initial locations (-8,-12,-20°), (-6,-18,-70°), (0,-10,0^o) and (8,-8,50^o). Notice that, in contrast to the use of the epipoles, the robot is able to reach the target from an initial location aligned without lateral error (0,-10,0^o).

In Fig. 4.6(b), both outputs (x and y positions) are driven to zero in 100 s and the robot reaches the target also with the desired orientation ($\phi = 0$). It is worth emphasizing that the previous behavior is obtained through the smooth control inputs shown in Fig. 4.6(c) for each initial location. These velocities start and end with zero value and they are always well defined. Although for these results the rotational velocity is more sensitive to the noise of the measurement, the resultant motion of the robot is not affected, as seen in the evolution of the robot state (Fig. 4.6(b)).

The results of the previous figures have been obtained for three different types of central cameras. Fig. 4.7(a) shows the motion of the image points for the case (-8,-12,-20°), in which a hypercatadioptric system is simulated. Fig. 4.7(b) corresponds to the case (-6,-18,-70°) with a paracatadioptric system and Fig. 4.7(c)

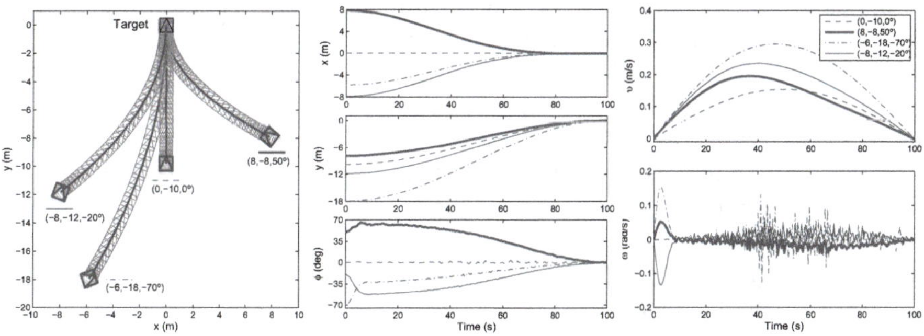

(a) (b) (c)

Fig. 4.6 Behavior of the robot motion as given by the smooth velocities obtained from the control law. (a) Paths on the $x-y$ plane. (b) State variables of the robot. (c) Computed velocities.

presents the points as seen for a conventional camera for the initial location (0,-10,0^o). Particularly in the last case, the 3D scene has been adapted to be in the field of view of the camera. This emphasize the advantage of using omnidirectional vision for visual servoing in order to avoid problems with the scene leaving the field of view. Additionally, the tensor estimation is in general more sensitive to noise for conventional images than for omnidirectional ones. A noisy estimated tensor yields a light heading that may become significative for conventional images, as can be seen in Fig. 4.7(c). For instance, the control for the last two initial locations is complicated using a conventional camera even by adapting the 3D scene, given that the large required rotation during the task makes the target to leave the field of view.

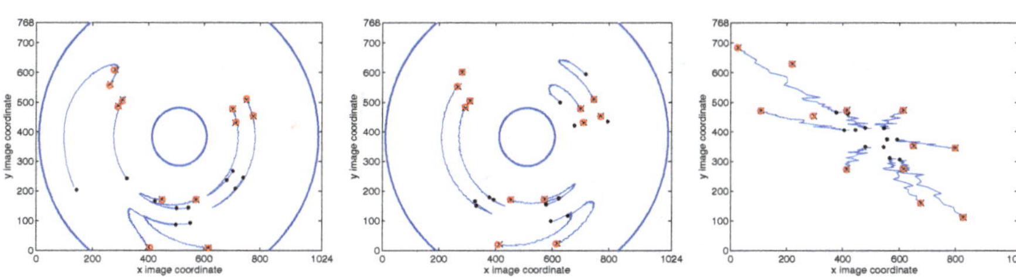

(a) (b) (c)

Fig. 4.7 Synthetic images showing the motion of the point features for (a) hypercatadioptric, (b) paracatadioptric and (c) conventional cameras. Each figure depicts the image points of the three views: initial image represented by black points, target image represented by red circles and image at the end of the motion represented by times signs.

Additionally, in order to test the performance in the servoing task, we have carried out Monte Carlo simulations. Table 4.1 shows that the target location (0,0,0^o) is always reached, and presents the average final location as obtained from 100 Monte Carlo runs. The final orientation is one degree or less with small standard deviation in any case. The robot reaches the target with neglected pose error for the initial location (0,-10,0^o) and it is not included in the table. The most challenging initial location in terms of lateral error is (8,-8,50^o), as can be seen in Fig. 4.6(a), and this is the reason of the largest e_x in the table. A similar behavior is observed regarding to e_y for the initial location (-6,-18,-70°). This large initial longitudinal error makes the largest final error for the $y-$coordinate. Nevertheless, these largest values correspond to errors that are small.

Table 4.1 Final error obtained by averaging 100 Monte Carlo runs to reach the target (0,0,0) from different initial locations

	(2m,-12m,45°)		(-6m,-16m,-20°)		(8m,-8m,50°)		(-6m,-18m,-70°)		(-8m,-12m,-20°)	
	Mean	St. dev.	Mean	St. dev.	Mean	St. dev.	Mean	St. dev.	Mean	St. dev.
e_x (cm)	0.90	1.03	-0.90	0.74	-6.4	0.2	3.5	0.5	-3.4	0.35
e_y (cm)	-0.54	0.18	-1.5	0.27	-1.3	0.2	-17.0	0.2	1.4	0.2
e_ϕ (deg)	-0.10	0.46	0.05	0.12	-1.0	0.13	-1.0	0.2	-0.5	0.15

4.4.2 Real-World Experiments

The control scheme of section 4.3 has been tested experimentally using the 1D trifocal tensor, given that this measurement provides better benefits than the epipolar geometry, mainly, avoidance of complete camera calibration and the problem of short baseline. The experimental platform used is presented in Fig. 1.1. The omnidirectional images are captured at a size of 800×600 pixels using the free software Player. The camera is connected to a laptop onboard the robot (Intel® Core^{TM} 2 Duo CPU at 2.50 GHz with operating system Debian Linux), in which the pose-estimation and the control law are implemented in C++. The observed scene has been set up with features on three different planes in order to ensure a sufficient number of points in the scene. However, points not belonging to these planes are also used to achieve a total of 13 points, which are manually matched in the three available images. These experiments have been carried out using a tracking of features as implemented in the OpenCV library. The tracking of features has been extensively applied for visual servoing purposes [97], it has a low computational cost and leads to a good behavior of the 1D trifocal tensor estimation.

The 1D trifocal tensor is estimated using the five-point method as described in section 3.2 with the projection center ($x_0 = 404$, $y_0 = 316$) as the only required information of the imaging system. The projection center has been previously estimated using a RANSAC approach from 3D vertical lines [59], which project in radial lines in central omnidirectional images. Thus, for this type of images, it is enough to find the point where radial lines join, which avoids to obtain the complete camera calibration parameters. The sampling time T_s is set to the same value as in the simulations (0.5 s). It is an adequate closed loop frequency that leads to a good behavior in the estimation of the 1D trifocal tensor. The distance from the camera to the rotation axis of the robot has been roughly set to $\ell = 10$ cm. The robot has been located in $x = -0.6$ m and $y = -1.8$ m from the target location and with the same orientation as the desired pose ($\phi = 0$). This metric information is considered as the ground truth for the experimental evaluation. The initial location is estimated through the initialization procedure of the dynamic estimation using the values of the elements of the 1D trifocal tensor (section 4.2.2.3).

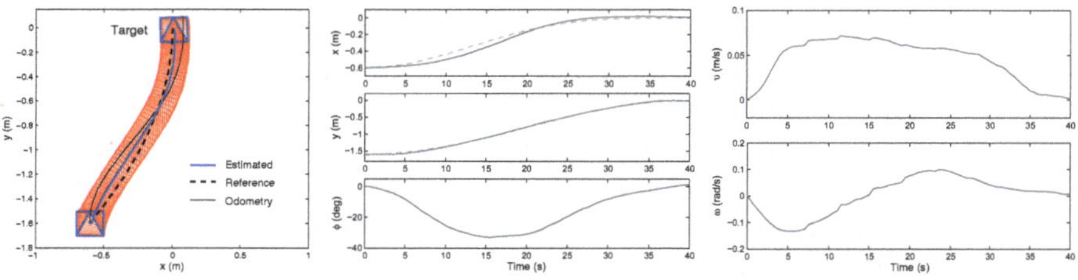

(a) (b) (c)

Fig. 4.8 Experimental results with the closed loop control system. (a) Resultant path. (b) Estimated camera-robot state. (c) Computed velocities. The path is plotted using the estimated camera-robot state, but also the reference path and the odometry are shown.

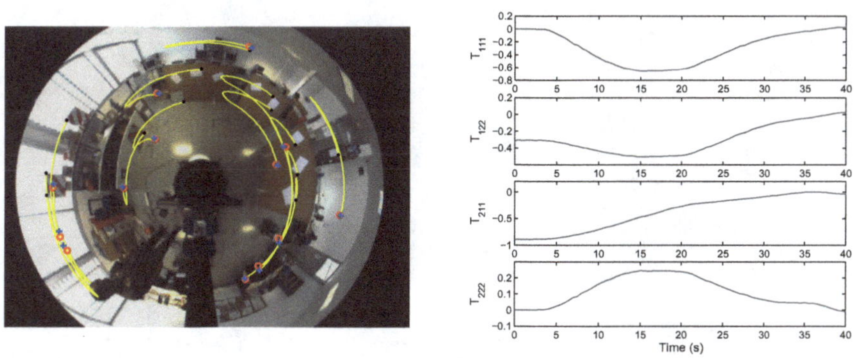

(a) (b)

Fig. 4.9 Behavior of the extracted information from the images for the real experiments. (a) Motion of the point features on the initial image. (b) Four normalized tensor elements. Regarding to the point features, the black points corresponds to the initial image, the red circles to the target image and the blue plus signs are the points in the image at the end of the motion.

Fig. 4.8(a) presents the resultant path, given by the estimated state of the robot, for one of the experimental runs. This figure also shows the reference path and the one given by odometry. It can be seen that the estimated path is closer to the reference than the path obtained from odometry. Thus, we assert that the estimation of the camera-robot pose is sufficiently accurate and then, the estimated pose is suitable for feedback control in the Cartesian space. The duration of the positioning task is fixed to 40 s through the time τ in the references, which is the termination condition of the control law. Fig. 4.8(b) shows the behavior of the estimated state together with the tracked references for the position coordinates. The performance of the reference tracking is better for the longitudinal than for the lateral position. It may be because the assumed simple robot model. Even so, the final position error is around 2 cm and the orientation error is practically negligible.

The input velocities as given by the control law using feedback of the estimated pose are shown in Fig. 4.8(c). They behave smoothly along the task, in contrast to the control inputs of previous approaches in the literature. Fig. 4.9(a) presents the motion of the image points along the sequence, where the points at the end of the motion (black plus signs) are almost superimposed to the points in the target image (red circles). Notice that the point features move smoothly, in such a way that the evolution of the tensor elements is also smooth during the task, as presented in Fig. 4.9(b). Also, it is worth noting that the tensor estimation is not affected when the robot is reaching the target, i.e., there is no problem with the short baseline. Fig. 4.10 shows a sequence of some images taken by the robot camera and an external video camera respectively.

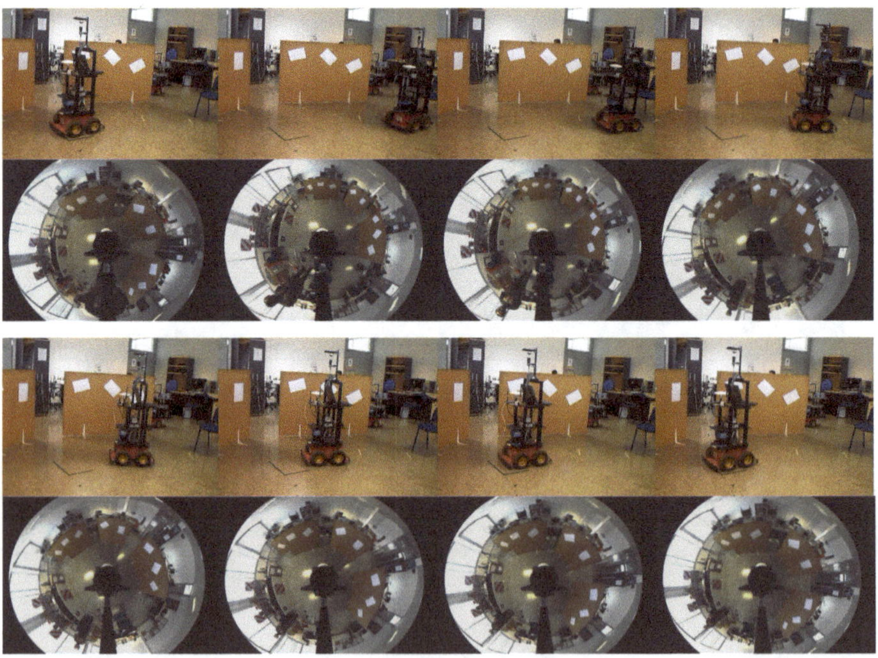

Fig. 4.10 Snapshots of the robot motion taken from an external camera for an experimental run and below, the omnidirectional images captured by the onboard hypercatadioptric camera for the corresponding snapshot's location. The first images in the first two rows correspond to the target location and the next images to the right correspond to the initial location. The last images in the last two rows correspond to the location at the end of the motion.

These results validate the effectiveness of the presented approach to reach a desired position and orientation using feedback of the estimated camera-robot pose from a geometric constraint. Notice that the estimation scheme may be generalized to 6 degrees of freedom by using a constant-velocity motion model for the system and some measurements from a geometric constraint. Thus, the visual servoing problem can be translated to the Cartesian space with the corresponding advantages given by the possibility to define a desired motion path and avoiding the visibility problem with omnidirectional vision.

4.5 Closure

In this chapter we have presented a new generic pose-estimation scheme that introduces a geometric constraint as measurement into a dynamic estimation process. Particularly, using the 1D trifocal tensor as measurement, this approach is a semicalibrated pose-estimation scheme. This approach does not need a target model neither scene reconstruction nor depth information. A novel comprehensive observability analysis using nonlinear tools has been presented, and the conditions to preserve observability depending on the velocities have been given. An additional benefit of exploiting the epipolar geometry or the 1D trifocal tensor for dynamic pose-estimation is the generality of the scheme, given that it is valid for any visual sensor obeying a central projection model. We have demonstrated the feasibility of closing a control loop using the estimated pose as feedback information in order to drive the robot to a desired location. Therefore, the described position-based control scheme solves the pose regulation problem avoiding visibility constraints by using omnidirectional vision. The control approach is a single step control law that corrects position and orientation simultaneously using smooth velocities. We have also shown the additional utility of feedback of the estimated pose when obstacles appear in the path during the execution of the task. The performance of this method is proved via simulations and real-world experiments using images from a hypercatadioptric imaging system.

Chapter 5
Conclusions

In this book we have presented and evaluated experimentally solutions to the problem of visual control of wheeled mobile robots using exclusively the information provided by an onboard monocular imaging system. The importance of addressing this problem is motivated by the increasing number of applications with this type of robots for service tasks. In this context, the general focus of the book is to give a formal treatment of the aspects from control theory applied in the particular problem of vision-based navigation of wheeled mobile robots, in such a way that vision and control have been unified to design control schemes with properties of stability, a large region of convergence (without local minima) and good robustness against parametric uncertainty and image noise.

Three different approaches have been presented along the book in order to address the pose regulation problem of mobile robots, where a desired position and orientation is defined to be reached. In this context, the control system provides suitable input velocities to drive the robot to the desired location using the teach-by-showing strategy. This implies that the target image must be previously known and the visual measurements are relative values between the locations associated to the target and current views. Thus, it is clear the need of a learning phase in a visual control loop, where the target image must be memorized. In pro of versatility, we have considered that no previous additional information about the scene is needed, i.e., any model of the environment and no artificial landmarks are used. Given that most of the mobile robots have non-holonomic motion constraints and they are underactuated systems that feature a degree of freedom in the robot dynamics, the presented control schemes are designed taking into account these properties.

In order to extract feedback information from the current and target images, it is required that both views share information, which means to have some common visual features in both images. In many cases, and especially when the initial camera position is far away from its desired value, the target features may leave the camera field of view during the navigation, which leads to failure because feedback error cannot be computed anymore. Recently, omnidirectional vision has attracted the attention of the robotics research community for the benefits provided by its wide field of view. This is motivated by the better understanding of those systems that

H.M. Becerra and C. Sagüés, *Visual Control of Wheeled Mobile Robots*,
Springer Tracts in Advanced Robotics 103,
DOI: 10.1007/978-3-319-05783-5_5, © Springer International Publishing Switzerland 2014

capture all the scene around a single view point, i.e., central imaging systems. In this sense, a general benefit of the control schemes presented in this book is that they are all valid for imaging systems obeying approximately a central projection model.

Because of the nonlinearity of the problem of visual control for wheeled mobile robots, singularities frequently appear when the robot velocities are computed using an input-output transformation of the system. The control systems described in this book cope with these singularities ensuring the stability of the feedback control system. There are many visual servoing schemes in the literature based on a pseudoinverse approach for non-square systems, which present potential problems of stability. We have designed square control systems where stability can be demonstrated with a large region of convergence and without local minima. Regarding to robustness, the effects of parametric uncertainty due to calibration errors and the effects of the measurement noise added to the feedback signals have been minimized through the use of the particular control technique sliding mode control.

The geometric constraints relate corresponding features in a multi-view framework and encapsulates their geometry in a few visual measurements. We have exploited these properties in order to have adequate task functions from a reduced set of measurements. Moreover, the geometric constraints provide a kind of filtering to the visual measurements. We focus on exploiting the epipolar geometry and the trifocal tensor given that they can be used for generic scenes, not limited to particular scenes like planar ones. For the epipolar constraint, we have taken advantage of the information provided by three images through their pairwise epipolar geometries. The pose regulation problem has been solved without needing to change to any approach other than epipolar-based control. The presented epipolar control deals with singularities induced by the epipolar geometry maintaining always bounded inputs, which allows the robot to carry out a direct motion toward the target.

Although the trifocal tensor is a geometric constraint that intrinsically integrates the rich information of three views, it has been little exploited for visual servoing. Moreover, this tensor is an improved visual measurement with respect to the epipolar geometry, more general, more robust and without the drawbacks of the epipolar geometry, like the problem of short baseline. We have exploited a simplified version of the trifocal tensor, the 1D trifocal tensor, for solving the pose regulation problem without commuting to any other approach. The 1D trifocal tensor is estimated from bearing information of the visual features, which allows exploiting the properties of omnidirectional images of preserving that information. The direct feedback of this tensor, for the stabilization of an adequate two-dimensional error function using a robust control law, results in a visual servoing scheme that requires only one calibration parameter of the imaging system.

As complement to the aforementioned image-based schemes and in order to reduce the dependence on the data from the image plane, we have presented an estimation scheme based also on the epipolar geometry and the trifocal tensor for position-based control purposes. It has been demonstrated that these geometric constraints can be used for dynamic pose-estimation by using a comprehensive nonlinear observability study. The presented method is the first semicalibrated pose-estimation scheme exploiting the properties of the 1D trifocal tensor. This approach does not need a target model neither scene reconstruction nor depth information. Additionally, we have demonstrated the feasibility of closing a control loop using the estimated pose as feedback information in order to drive the robot to a desired location. Therefore, this position-based control approach solves the pose regulation problem avoiding visibility constraints by means of omnidirectional vision. The control approach is a single-step control law that corrects the robot pose using smooth velocities.

The different control schemes presented along this book have been validated experimentally. Simulations and real-world experiments with different platforms and imaging systems have been carried out to show the performance of the approaches. The real-world experimentation has shown that all the described control schemes are feasible to be implemented in a common laptop, providing adequate closed loop frequency that can be achieved with typical experimental hardware. There are several open issues in the problem of visual control, however, as shown in this book, the use of a geometric constraint as visual measurements and adequate control techniques gives improved performance for positioning tasks, which can be extended, for instance, to long distance navigation of wheeled mobile robots and also can be explored for vision-based control of humanoid robots.

Appendix A
Basics of Nonlinear Control and State Estimation

Abstract. Control theory is a strong and rigorous field of knowledge by itself and its application is well appreciated in many other fields including robotics. In some cases, visual control has not taken the maximum advantage of this theory. Along the book, we make an effort to treat all the problems from a control theory point of view. In this appendix, we describe briefly two important control techniques that are used in the visual servoing schemes presented in the book and we introduce important concepts about estimation of the state of nonlinear systems.

A.1 Input-Output Linearization

This control technique is also known as exact feedback linearization and is based on the possibility of nonlinearities cancellation. There are structural properties of the systems that allow us to perform such cancellation as described in this section. This control technique is the basis of all the control laws presented in the book and it is referred to in each chapter. For more details about this control technique refer to [124], [70], [129], [72]. Consider the single-input-single-output (SISO) control system

$$\begin{aligned} \dot{\mathbf{x}} &= f(\mathbf{x}) + g(\mathbf{x})u, \\ y &= h(\mathbf{x}), \end{aligned}$$

where the state vector $\mathbf{x} \in \mathbb{R}^n$, the control vector $u \in \mathbb{R}$, the output vector $y \in \mathbb{R}$, and the vector fields $f(\mathbf{x}) \in \mathbb{R}^n$ and $g(\mathbf{x}) \in \mathbb{R}^{n \times 1}$. The condition for a system to be input-output linearizable consists in finding a function $T_1(\mathbf{x})$ for every output that satisfies

$$\frac{\partial T_i}{\partial \mathbf{x}} g(\mathbf{x}) = 0, \quad i = 1, 2, ..., n-1; \quad \frac{\partial T_n}{\partial \mathbf{x}} g(\mathbf{x}) \neq 0, \tag{A.1}$$

with

$$T_{i+1}(\mathbf{x}) = \frac{\partial T_i}{\partial \mathbf{x}} f(\mathbf{x}), \quad i = 1, 2, ..., n-1.$$

This condition can be interpreted as a restriction on the way the derivatives of y depend on u, which is defined by the structure of the system. To see this point, set $\psi_1(\mathbf{x}) = h(\mathbf{x})$. The derivative $\dot{y}$ is given by

$$\dot{y} = \frac{\partial \psi_1}{\partial \mathbf{x}} \left[f(\mathbf{x}) + g(\mathbf{x})u \right].$$

If $[\partial \psi_1 / \partial \mathbf{x}] g(\mathbf{x}) = 0$, then

$$\dot{y} = \frac{\partial \psi_1}{\partial \mathbf{x}} f(\mathbf{x}) = \psi_2(\mathbf{x}).$$

If we continue to calculate the second derivative of y, denoted by $y^{(2)}$, we obtain

$$y^{(2)} = \frac{\partial \psi_2}{\partial \mathbf{x}} \left[f(\mathbf{x}) + g(\mathbf{x})u \right].$$

Once again, if $[\partial \psi_2 / \partial \mathbf{x}] g(\mathbf{x}) = 0$, then

$$y^{(2)} = \frac{\partial \psi_2}{\partial \mathbf{x}} f(\mathbf{x}) = \psi_3(\mathbf{x}).$$

Repeating this process, we see that if $h(\mathbf{x}) = \psi_1(\mathbf{x})$ satisfies (A.1); that is,

$$\frac{\partial \psi_i}{\partial \mathbf{x}} g(\mathbf{x}) = 0, \quad i = 1, 2, ..., n-1; \quad \frac{\partial \psi_n}{\partial \mathbf{x}} g(\mathbf{x}) \neq 0,$$

where

$$\psi_{i+1}(\mathbf{x}) = \frac{\partial \psi_i}{\partial \mathbf{x}} f(\mathbf{x}), \quad i = 1, 2, ..., n-1.$$

Then the control input u does not appear in the equations of the first $n-1$ derivatives $y, \dot{y}, ..., y^{(n-1)}$ and appears in the equation of $y^{(n)}$ with a nonzero coefficient

$$y^{(n)} = \frac{\partial \psi_n}{\partial \mathbf{x}} f(\mathbf{x}) + \frac{\partial \psi_n}{\partial \mathbf{x}} g(\mathbf{x})u.$$

This equation shows clearly that the system is input-output linearizable since the state feedback control with auxiliary input υ

$$u = \frac{1}{\frac{\partial \psi_n}{\partial \mathbf{x}} g(\mathbf{x})} \left[-\frac{\partial \psi_n}{\partial \mathbf{x}} f(\mathbf{x}) + \upsilon \right], \tag{A.2}$$

reduces the input-output map to $y^{(n)} = \upsilon$, which is a chain of n integrators, i.e., a linear system whose new dynamics can be assigned by the auxiliary input.

There exist cases where the control input u appears in the equation of one of the derivatives $y, \dot{y}, ..., y^{(n-1)}$. If the coefficient of u (when it happens) is nonzero, then we can again linearize the input-output map. In particular, if $h = \psi_1(\mathbf{x})$ satisfies

$$\frac{\partial \psi_i}{\partial \mathbf{x}} g(\mathbf{x}) = 0, \quad i = 1, 2, ..., r-1; \quad \frac{\partial \psi_r}{\partial \mathbf{x}} g(\mathbf{x}) \neq 0$$

for some $1 \le r < n$, then the equation of $y^{(r)}$ is given by

$$y^{(r)} = \frac{\partial \psi_r}{\partial \mathbf{x}} f(\mathbf{x}) + \frac{\partial \psi_r}{\partial \mathbf{x}} g(\mathbf{x}) u.$$

Therefore, the state feedback control with auxiliary input υ

$$u = \frac{1}{\frac{\partial \psi_r}{\partial \mathbf{x}} g(\mathbf{x})} \left[-\frac{\partial \psi_r}{\partial \mathbf{x}} f(\mathbf{x}) + \upsilon \right], \tag{A.3}$$

linearizes the input-output map to the chain of r integrators $y^{(r)} = \upsilon$. In this case, the integer r is called the relative degree of the system. This allows to enunciate the following definition that can be found in [124].

Definition A.1. The internal unobservable dynamics that results in the system when $r < n$ is called *zero dynamics*. This dynamics is described by a subset of the state space which makes the output to be identically zero. Keeping the output identically zero implies that the solution of the state equation must be confined to the set

$$Z^* = \{\mathbf{x} \in \mathbb{R}^n \mid \psi_1(\mathbf{x}) = \psi_2(\mathbf{x}) = ... = \psi_r(\mathbf{x}) = 0\}. \tag{A.4}$$

This control technique is directly extendable to multi-input-multi-output (MIMO) square systems. The previous process that has been shown must be repeated for each one of the m outputs for a MIMO system. The process verifies the appearance of any of the control inputs in the equation of one of the output derivatives. Thus, the linearized dynamics of the system results in m blocks of $r_1, ..., r_m$ integrators. The vector relative degree $\{r_1, ..., r_m\}$ establishes the occurrence of a zero dynamics if $r_1 + r_2 + ... + r_m < n$.

Notice that the cancellation of nonlinearities depends on the exact knowledge of the model of the system. This is the reason because the input-output linearization is not a robust control technique. It is common to have uncertainty in some parameters of the systems, which may degenerate the controller performance and the property of stability of the closed loop.

A.2 A Robust Control Technique: Sliding Mode Control

Sliding mode control is gaining importance as a universal design methodology for the robust control of linear and nonlinear systems. It offers several interesting features from a control theory point of view [138], namely, an inherent order reduction, direct incorporation of robustness against system uncertainties and disturbances, and an implicit stability proof. In spite of these kind benefits, sliding mode control has been little exploited in visual servoing, however, its good performance can be verified in the control schemes of chapter 2 and chapter 3.

Basically, sliding mode control makes use of a high-speed switching control law to drive the nonlinear state trajectory of the system onto a specified and user-chosen surface in the state space (called the sliding or switching surface) in a finite time, and to keep the state trajectory on this surface for all subsequent time. The plant dynamics constrained to this surface represent the controlled system's behavior, which is independent of matched uncertainties and disturbances. Matched uncertainties are those that belong to the range space of the input vector. This is the so-called *matching condition*. By proper design of the sliding surface, sliding mode control achieves the conventional goals of control such as stabilization, tracking, regulation, etc.

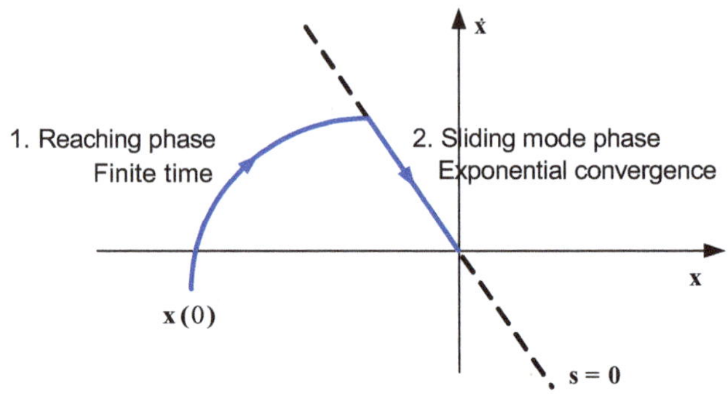

Fig. A.1 Phase portrait of sliding modes control showing the two phases of the control

Sliding modes are well studied in a class of systems having a state model nonlinear in the state vector $\mathbf{x}$ and linear in the control vector $\mathbf{u}$ (affine systems) of the form $\dot{\mathbf{x}} = f(t,\mathbf{x},\mathbf{u}) = f(t,\mathbf{x}) + g(t,\mathbf{x})\mathbf{u}$, which is accomplished by the camera-robot model (1.2) with the particularity that $f(t,\mathbf{x}) = \mathbf{0}$.

The *switching surface* is so called because if the state trajectory of the system is "above" the surface, then the control input has one gain, and a different gain is applied if the trajectory drops "below" the surface. Thus, each entry $u_i(t)$ of the switched control $\mathbf{u}(t) \in \mathbb{R}^m$ has the form

$$u_i(t,\mathbf{x}) = \begin{Bmatrix} u_i^+(t,\mathbf{x}) & \text{if } s_i(\mathbf{x}) > 0 \\ u_i^-(t,\mathbf{x}) & \text{if } s_i(\mathbf{x}) < 0 \end{Bmatrix}, i = 1,\ldots,m, \tag{A.5}$$

where $s_i = 0$ is the i-th switching surface associated with the $(n-m)$-dimensional switching surface

$$\mathbf{s}(\mathbf{x}) = \begin{bmatrix} s_1(\mathbf{x}) & \cdots & s_m(\mathbf{x}) \end{bmatrix}^T = \mathbf{0}.$$

Sliding mode control design breaks down into two stages as shown in Fig. A.1. Stage one implies constructing switching surfaces so that the system restricted to these surfaces produces a desired behavior. Stage two involves constructing switched feedback gains which drive the state trajectory to the sliding surface maintaining it there. Then, the action of a sliding mode control law is performed in two phases, a reaching phase, during which, trajectories starting off the surface $\mathbf{s} = \mathbf{0}$ move toward it and reach it in a finite time, followed by a sliding phase, during which, the motion is confined to the surface $\mathbf{s} = \mathbf{0}$ and the dynamics of the system are represented by a reduced order model.

An undesirable phenomenon presented in sliding mode control systems is the chattering. This is an oscillation within a neighborhood of the switching surface such that $\mathbf{s} = \mathbf{0}$ is not satisfied for all time after the switching surface is reached. If the frequency of the switching is very high compared with the dynamic response of the system the chattering problem is often although not always negligible.

A.3 Theory of State Observability

An important concept that is related to the implementation of feedback control laws is the observability, which allows to verify whether the state variables can be uniquely determined from a given set of constraints (output measurements). Observability is a structural property that may affect the convergence of an estimation scheme. This property specifies if two states are distinguishable by measuring the output, i.e.,

$$\mathbf{x}_1 \neq \mathbf{x}_2 \Longrightarrow h(\mathbf{x}_1) \neq h(\mathbf{x}_2).$$

Intuitively, this means that if two measurements from the sensors are different, then the states are also different [65]. There are few works concerned about the state observability for mobile robots. On one hand, some of them take advantage of linearized models using the theory of linear systems to analyze the observability of the SLAM problem [139], [26]. On the other hand, there also exist some contributions where a nonlinear observability analysis is carried out for localization [102] or for SLAM applications [77], [68]. In the rest of the section, we introduce the theory to analyze the observability with emphasis on nonlinear systems as merit the treatment of our visual servoing problem, especially in chapter 4.

A.3.1 Nonlinear Continuous Systems

In the scope of this book, we deal with a nonlinear estimation problem and the observability analysis of this case requires different tools unlike the linear case. The theory for the observability analysis of continuous systems has been introduced in [65]. According to this theory, the following *observability rank condition* can be enunciated for a general affine system.

Definition A.2. The continuous-time nonlinear system $\dot{\mathbf{x}} = \begin{bmatrix} g_1(\mathbf{x}) & g_2(\mathbf{x}) & \cdots & g_m(\mathbf{x}) \end{bmatrix} \mathbf{u}$ with a measurement vector $h(\mathbf{x}) \in \mathbb{R}^p$ is locally weakly observable if the observability matrix with rows

$$\mathbf{O} \triangleq \left[\nabla L^q_{g_i g_j} h_k(\mathbf{x}) \mid i,j = 1,2,...,m,\ k = 1,2,...,p,\ q \in \mathbb{N} \right]^T \tag{A.6}$$

is of full rank n.

The expression $L^q_{g_i} h_k(\mathbf{x})$ denotes the qth order Lie derivative of the scalar function h_k along the vector field g_i. Thus, the matrix (A.6) is formed by the gradient vectors

$\nabla L^{q}_{g_i g_j} h_k(\mathbf{x})$ that span a space containing all the possible Lie derivatives. From the previous definition, we can see that the observability property of the camera-robot system depends completely on the excitations, given that it is a driftless system.

Considering that the vector fields g_i and the measurement functions h_k are infinitely smooth, the matrix (A.6) could have infinite number of rows. However, it suffices to find a set of rows linearly independent in order to fulfill the rank condition. Locally weak observability is a concept stronger than observability, which states that one can instantaneously distinguish each point of the state space from its neighbors, without necessity to travel a considerable distance, as admitted by the observability concept.

A.3.2 Nonlinear Discrete Systems

Actually, the implementation of any estimation and control algorithm must be performed in discrete time. Consequently, it is necessary to analyze the properties of the camera-robot system in its discrete representation. To do that, the following observability rank condition for discrete nonlinear systems is used [112].

Definition A.3. The generic discrete-time nonlinear system given by (1.5) is said to be observable if the nonlinear observability matrix

$$\mathbf{O}_d = \begin{bmatrix} \mathbf{H}_k \\ \mathbf{H}_{k+1}\mathbf{F}_k \\ \vdots \\ \mathbf{H}_{k+n-1}\mathbf{F}_{k+n-2}\cdots\mathbf{F}_k \end{bmatrix} = \begin{bmatrix} \frac{\partial h}{\partial \mathbf{x}}(\mathbf{x}_k) \\ \frac{\partial h}{\partial \mathbf{x}}(\mathbf{x}_{k+1})\frac{\partial f}{\partial \mathbf{x}}(\mathbf{x}_k) \\ \vdots \\ \frac{\partial h}{\partial \mathbf{x}}(\mathbf{x}_{k+n-1})\frac{\partial f}{\partial \mathbf{x}}(\mathbf{x}_{k+n-2})\cdots\frac{\partial f}{\partial \mathbf{x}}(\mathbf{x}_k) \end{bmatrix} \quad \text{(A.7)}$$

is of full rank n.

This matrix resembles the observability matrix for linear time invariant systems in appearance, but is different in that the matrices of the linear approximation are not constant. This observability rank condition allows to conclude if the estimator collects enough information along $n-1$ instant times in order to make the system observable

A.3.3 Discrete Piece-Wise Constant Systems

A nonlinear system can be approximated by a piece-wise constant system using the linearization $\mathbf{F}_k = \partial f/\partial \mathbf{x}_k$, $\mathbf{H}_k = \partial h/\partial \mathbf{x}_k$ for each instant time k. According to the theory in [57], the observability of piece-wise constant system can be studied using the so-called stripped observability matrix for a number $r \in \mathbb{N}$ of instant times. We can summarize this theory in the following definition.

Definition A.4. Let the stripped observability matrix (SOM) be defined from the local observability matrices $\mathbf{O}_k$ as follows

$$\mathbf{O}_{SOM,p} = \begin{bmatrix} \mathbf{O}_k \\ \mathbf{O}_{k+1} \\ \vdots \\ \mathbf{O}_{k+p} \end{bmatrix} \text{ with } \mathbf{O}_k = \begin{bmatrix} \mathbf{H}_k \\ \mathbf{H}_k\mathbf{F}_k \\ \vdots \\ \mathbf{H}_k\mathbf{F}_k^{n-1} \end{bmatrix}. \tag{A.8}$$

Whether it is satisfied that $\mathbf{F}_k\mathbf{x}_k = \mathbf{x}_k \ \forall \ \mathbf{x}_k \in NULL(\mathbf{O}_k)$, then the discrete piecewise constant system is completely observable if and only if $\mathbf{O}_{SOM,p}$ is of rank n.

This claims that observability can be gained in some steps r even if local observability is not ensured.

A.4 Dynamic Pose Estimation

In the framework of this book, a good option to deal with the problem of pose-estimation for feedback control is to tackle it as a *nonlinear dynamic estimation problem*. In the literature, the pose-estimation problem for visual servoing purposes has also been treated through static approaches [24], [51], [119], [43]. These approaches are based on the homography or the essential matrix decomposition at each instant time, which results computationally costly. In contrast, the dynamic approach can be solved through an efficient algorithm as the extended Kalman filter (EKF). This filter gives an approximation of the optimal estimate. The nonlinearities of the systems's dynamics are approximated by a linearized version of the nonlinear system model around the last state estimate.

The EKF is an scheme widely used to estimate the state of nonlinear observable systems. Particularly, it has been applied previously in the visual servoing problem [142], [45], [127]. This approach is a solution to the estimation problem that provides generality to a visual servoing scheme in comparison to nonlinear observers that are designed for a particular system. Moreover, the EKF offers advantages over other estimation methods, e.g., temporal filtering, recursive implementation or ability to change the set of measurements during the operation if required, etc. Also, the pose prediction computed by the filter may be used to set up a dynamic windowing technique for the search of image features. Moreover, the basic form of the EKF provides a good compromise between accuracy in the estimation and computational cost.

Consider the discrete-time dynamic system (1.5) that describes the robot-camera kinematics and models unpredictable disturbances through additive Gaussian noises. In this framework, the prediction equations to compute the estimated state are

$$\begin{aligned} \hat{\mathbf{x}}_k^- &= f\left(\hat{\mathbf{x}}_{k-1}^+, \mathbf{u}_{k-1}\right), \\ \mathbf{P}_k^- &= \mathbf{F}_k\mathbf{P}_{k-1}^+\mathbf{F}_k^T + \mathbf{G}_k\mathbf{M}_{k-1}\mathbf{G}_k^T, \end{aligned} \tag{A.9}$$

where the linear approximation $\hat{\mathbf{x}}_{k+1} = \mathbf{F}_k\hat{\mathbf{x}}_k + \mathbf{G}_k\mathbf{u}_k$, $\hat{\mathbf{y}}_k = \mathbf{H}_k\hat{\mathbf{x}}_k$ of the nonlinear system is used. The update equations to correct the estimates are

$$\begin{aligned}\mathbf{Q}_k &= \mathbf{H}_k \mathbf{P}_k^- \mathbf{H}_k^T + \mathbf{N}_k,\\ \mathbf{K}_k &= \mathbf{P}_k^- \mathbf{H}_k^T (\mathbf{Q}_k)^{-1},\\ \hat{\mathbf{x}}_k^+ &= \hat{\mathbf{x}}_k^- + \mathbf{K}_k \nu_k,\\ \mathbf{P}_k^+ &= [\mathbf{I} - \mathbf{K}_k \mathbf{H}_k] \mathbf{P}_k^- .\end{aligned} \tag{A.10}$$

In these equations, $\hat{\mathbf{x}}_k^-$, $\mathbf{P}_k^-$ represent an *a priori* estimate of the state and its covariance, and $\hat{\mathbf{x}}_k^+$, $\mathbf{P}_k^+$ provide an *a posteriori* estimated state for step k. It means that the a posteriori information utilizes feedback error in order to improve the state estimation. The measurement innovation $\nu_k = \mathbf{y}_k - h\left(\hat{\mathbf{x}}_k^-\right)$ and its covariance matrix $\mathbf{Q}_k$ are also used to verify the consistency property of the estimation in real situation. A typical statistical test of a Kalman estimator is the consistency test. It determines if the computed covariances will match the actual estimation errors [10]. A consistency index can be defined as $CI = D^2/\chi^2_{n,1-\alpha}$, where D^2 is the normalized estimate error square (NEES) or the normalized innovation squared (NIS), n is the dimension of the vector state or the dimension of the measurement and $(1-\alpha)$ is the confidence level (95% typically) in the chi-square distribution. The NEES is computed from the estimation error as

$$NEES_k = \tilde{\mathbf{x}}_k^T \left(\mathbf{P}_k^+\right)^{-1} \tilde{\mathbf{x}}_k, \tag{A.11}$$

where $\tilde{\mathbf{x}}_k = \mathbf{x}_k - \hat{\mathbf{x}}_k$.The NIS is computed from the measurement innovation ν_k as follows

$$NIS_k = \nu_k^T (\mathbf{Q}_k)^{-1} \nu_k. \tag{A.12}$$

In any case, when $CI < 1$, the estimation is consistent, otherwise it is optimistic or inconsistent and the estimation may diverge. It can happen if the consecutive linearization is not a good approximation of the nonlinear model in all the associated uncertainty domain. Next, we present the required Jacobian matrices in the linear approximation of the discrete system (1.4)

$$\begin{aligned}\mathbf{F}_k &= \left|\frac{\partial f}{\partial \mathbf{x}_k}\right|_{\mathbf{x}_k = \hat{\mathbf{x}}_k^+, \mathbf{m}_k = 0} = \begin{bmatrix} 1 & 0 & \Delta_{y,k} \\ 0 & 1 & -\Delta_{x,k} \\ 0 & 0 & 1 \end{bmatrix}_{\phi_k = \hat{\phi}_k^+}, \\ \mathbf{G}_k &= \left|\frac{\partial f}{\partial \mathbf{u}_k}\right|_{\mathbf{x}_k = \hat{\mathbf{x}}_k^+} = \begin{bmatrix} -\sin\phi_k & -\ell\cos\phi_k \\ \cos\phi_k & -\ell\sin\phi_k \\ 0 & 1 \end{bmatrix}_{\phi_k = \hat{\phi}_k^+}, \\ \mathbf{H}_k &= \left|\frac{\partial h}{\partial \mathbf{x}_k}\right|_{\mathbf{x}_k = \hat{\mathbf{x}}_k^-, \mathbf{n}_k = 0},\end{aligned} \tag{A.13}$$

where $\Delta_{x,k} = T_s\left(\omega_k \ell \cos\phi_k + \upsilon_k \sin\phi_k\right)$, $\Delta_{y,k} = T_s\left(\omega_k \ell \sin\phi_k - \upsilon_k \cos\phi_k\right)$. Since $\mathbf{H}_k$ depends on the measurement model, it will be defined as needed for the estimation from different visual data.

References

1. Intel's OpenCV, http://sourceforge.net/projects/opencvlibrary/
2. Abdelkader, H.H., Mezouar, Y., Andreff, N., Martinet, P.: 2-1/2 D visual servoing with central catadioptric cameras. In: IEEE/RSJ International Conference on Intelligent Robots and Systems, pp. 2342–2347 (2005)
3. Abdelkader, H.H., Mezouar, Y., Andreff, N., Martinet, P.: Image-based control of mobile robot with central catadioptric cameras. In: IEEE International Conference on Robotics and Automation, pp. 3522–3527 (2005)
4. Abdelkader, H.H., Mezouar, Y., Martinet, P.: Path planning for image based control with omnidirectional cameras. In: IEEE Conference on Decision and Control, pp. 1764–1769 (2006)
5. Abdelkader, H.H., Mezouar, Y., Martinet, P., Chaumette, F.: Catadioptric visual servoing from 3D straight lines. IEEE Transactions on Robotics 24(3), 652–665 (2008)
6. Aicardi, M., Casalino, G., Bicchi, A., Balestrino, A.: Closed loop steering of unicycle-like vehicles via Lyapunov techniques. IEEE Robotics and Automation Magazine 2(1), 27–35 (1995)
7. Andreff, N., Espiau, B., Horaud, R.: Visual servoing from lines. The International Journal of Robotics Research 21(8), 679–699 (2002)
8. Atassi, A.N., Khalil, H.K.: A separation principle for the stabilization of a class of nonlinear systems. IEEE Transactions on Automatic Control 44(9), 1672–1687 (1999)
9. Baker, S., Nayar, S.K.: A theory of single-viewpoint catadioptric image formation. International Journal of Computer Vision 35(2), 175–196 (1999)
10. Bar-Shalom, Y., Li, X.R., Kirubarajan, T.: Estimation with Applications to Tracking and Navigation. John Willey and Sons, New York (2001)
11. Barreto, J.P., Araujo, H.: Geometric properties of central catadioptric line images and their application in calibration. IEEE Transactions on Pattern Analysis and Machine Intelligence 27(8), 1327–1333 (2005)
12. Barreto, J.P., Martin, F., Horaud, R.: Visual servoing/tracking using central catadioptric images. In: Siciliano, B., Dario, P. (eds.) Experimental Robotics VIII. STAR, vol. 5, pp. 245–254. Springer, Heidelberg (2002)
13. Basri, R., Rivlin, E., Shimshoni, I.: Visual homing: Surfing on the epipoles. International Journal of Computer Vision 33(2), 117–137 (1999)
14. Becerra, H.M., López-Nicolás, G., Sagüés, C.: Omnidirectional visual control of mobile robots based on the 1D trifocal tensor. Robotics and Autonomous Systems 58(6), 796–808 (2010)

15. Becerra, H.M., López-Nicolás, G., Sagüés, C.: A sliding mode control law for mobile robots based on epipolar visual servoing from three views. IEEE Transactions on Robotics 27(1), 175–183 (2011)
16. Becerra, H.M., Sagüés, C.: Sliding mode control for visual servoing of mobile robots using a generic camera. In: Bartoszewicz, A. (ed.) Sliding Mode Control, ch. 12, pp. 221–236. InTech, Vienna (2011)
17. Becerra, H.M., Sagüés, C.: A sliding mode control law for epipolar visual servoing of differential-drive robots. In: IEEE/RSJ International Conference on Intelligent Robots and Systems, pp. 3058–3063 (2008)
18. Becerra, H.M., Sagüés, C.: A novel 1D trifocal-tensor-based control for differential-drive robots. In: IEEE International Conference on Robotics and Automation, pp. 1104–1109 (2009)
19. Becerra, H.M., Sagüés, C.: Pose-estimation-based visual servoing for differential-drive robots using the 1D trifocal tensor. In: IEEE/RSJ International Conference on Intelligent Robots and Systems, pp. 5942–5947 (2009)
20. Becerra, H.M., Sagüés, C.: Dynamic pose-estimation from the epipolar geometry for visual servoing of mobile robots. In: IEEE International Conference on Robotics and Automation, pp. 417–422 (2011)
21. Becerra, H.M., Sagüés, C.: Exploiting the trifocal tensor in dynamic pose-estimation for visual control. IEEE Transactions on Control Systems Technology 21(5), 1931–1939 (2013)
22. Benhimane, S., Malis, E.: A new approach to vision-based robot control with omni-directional cameras. In: IEEE International Conference on Robotics and Automation, pp. 526–531 (2006)
23. Benhimane, S., Malis, E.: Homography-based 2D visual tracking and servoing. International Journal of Robotics Research 26(7), 661–676 (2007)
24. Benhimane, S., Malis, E., Rives, P., Azinheira, J.R.: Vision-based control for car platooning using homography decomposition. In: IEEE International Conference on Robotics and Automation, pp. 2173–2178 (2005)
25. Brockett, R.W.: Asymptotic stability and feedback stabilization. In: Brockett, R.W., Millman, R.S., Sussmann, H.J. (eds.) Differential Geometric Control Theory, pp. 181–191. Birkhäuser, Boston (1983)
26. Bryson, M., Sukkarieh, S.: Observability analysis and active control for airbone SLAM. IEEE Transactions on Aerospace and Electronic Systems 44(1), 261–280 (2008)
27. Cervera, E., del Pobil, A.P., Berry, F., Martinet, P.: Improving image-based visual servoing with three-dimensional features. The International Journal of Robotics Research 22(10-11), 821–839 (2003)
28. Hager, G.D., Chang, W.C., Morse, A.S.: Robot hand-eye coordination based on stereo vision. IEEE Control Systems Magazine 15(1), 30–39 (1995)
29. Chaumette, F.: Potential problems of stability and convergence in image-based and position-based visual servoing. In: Kriegman, D.J., Hager, G.D., Morse, A.S. (eds.) The Confluence of Vision and Control. LNCIS, vol. 237, pp. 66–78. Springer, Heidelberg (1998)
30. Chaumette, F.: Image moments: A general and useful set of features for visual servoing. IEEE Transactions on Robotics 20(4), 713–723 (2004)
31. Chaumette, F., Hutchinson, S.: Visual servo control. I. Basic approaches. IEEE Robotics and Automation Magazine 13(4), 82–90 (2006)
32. Chaumette, F., Hutchinson, S.: Visual servo control part II: Advanced approaches. IEEE Robotics and Automation Magazine 14(1), 109–118 (2007)

33. Chaumette, F., Rives, P., Espiau, B.: Positioning of a robot with respect to an object, tracking it and estimating its velocity by visual servoing. In: IEEE Intemational Conference on Robotics and Automation, pp. 2248–2253 (1991)
34. Chen, J., Dawson, D.M., Dixon, W.E., Behal, A.: Adaptive homography-based visual servo tracking for a fixed camera configuration with a camera-in-hand extension. IEEE Transactions on Control Systems Technology 13(5), 814–825 (2005)
35. Chen, J., Dawson, D.M., Dixon, W.E., Chitrakaran, V.K.: Navigation function-based visual servo control. Automatica 43(1), 1165–1177 (2007)
36. Chesi, G., Hashimoto, K., Prattichizzo, D., Vicino, A.: Keeping features in the field of view in eye-in-hand visual servoing: A switching approach. IEEE Transactions on Robotics 20(5), 908–913 (2004)
37. Chesi, G., Hung, Y.S.: Global path-planning for constrained and optimal visual servoing. IEEE Transactions on Robotics 23(5), 1050–1060 (2007)
38. Comport, A.I., Malis, E., Rives, P.: Real-time quadrifocal visual odometry. The International Journal of Robotics Research 29(2-3), 245–266 (2010)
39. Conticelli, F., Allotta, B., Khosla, P.K.: Image-based visual servoing of nonholonomic mobile robots. In: IEEE Conference on Decision and Control, pp. 3496–3501 (1999)
40. Corke, P.I., Hutchinson, S.A.: A new partitioned approach to image-based visual servo control. IEEE Transactions on Robotics and Automation 17(4), 507–515 (2001)
41. Courbon, J., Mezouar, Y., Eck, L., Martinet, P.: A generic fisheye camera model for robotics applications. In: IEEE International Conference on Intelligent Robots and Systems, pp. 1683–1688 (2007)
42. Courbon, J., Mezouar, Y., Martinet, P.: Indoor navigation of a non-holonomic mobile robot using a visual memory. Autonomous Robots 25(3), 253–266 (2008)
43. Courbon, J., Mezouar, Y., Martinet, P.: Autonomous navigation of vehicles from a visual memory using a generic camera model. IEEE Transactions on Intelligent Transportation Systems 10(3), 392–402 (2009)
44. Cretual, A., Chaumette, F.: Visual servoing based on image motion. The International Journal of Robotics Research 20(11), 857–877 (2001)
45. Das, A.K., Fierro, R., Kumar, V., Southall, B., Spletzer, J., Taylor, C.J.: Real-time vision-based control of a nonholonomic mobile robot. In: IEEE/RSJ International Conference on Intelligent Robots and Systems, pp. 1714–1718 (2001)
46. Davison, A.J., Murray, D.W.: Simultaneous localization and map-building using active vision. IEEE Transactions on Pattern Analysis and Machine Intelligence 24(7), 865–880 (2002)
47. de Wit, C.C., Sordalen, O.J.: Exponential stabilization of mobile robots with nonholonomic constraints. IEEE Transactions on Automatic Control 37(11), 1791–1797 (1992)
48. Deguchi, K.: Optimal motion control for image-based visual servoing by decoupling translation and rotation. In: IEEE/RSJ International Conference on Intelligent Robots and Systems, pp. 705–711 (1998)
49. Dellaert, F., Stroupe, A.W.: Linear 2D localization and mapping for single and multiple robot scenarios. In: IEEE International Conference on Robotics and Automation, pp. 688–694 (2002)
50. Espiau, B., Chaumette, F., Rives, P.: A new approach to visual servoing in robotics. IEEE Transactions on Robotics and Automation 8(3), 313–326 (1992)
51. Fang, Y., Dixon, W.E., Dawson, D.M., Chawda, P.: Homography-based visual servo regulation of mobile robots. IEEE Transactions on Systems, Man, and Cybernetics - Part B: Cybernetics 35(5), 1041–1050 (2005)
52. Fontanelli, D., Danesi, A., Belo, F.A.W., Salaris, P., Bicchi, A.: Visual servoing in the large. The International Journal of Robotics Research 28(6), 802–814 (2007)

53. Gans, N.R., Hutchinson, S.A.: A stable vision-based control scheme for nonholonomic vehicles to keep a landmark in the field of view. In: IEEE International Conference on Robotics and Automation, pp. 2196–2200 (2007)
54. Gans, N.R., Hutchinson, S.A.: Stable visual servoing through hybrid switched-system control. IEEE Transactions on Robotics 23(3), 530–540 (2007)
55. Geyer, C., Daniilidis, K.: Catadioptric projective geometry. International Journal of Computer Vision 16(6), 890–898 (2000)
56. Geyer, C., Daniilidis, K.: An unifying theory for central panoramic systems and practical implications. In: Vernon, D. (ed.) ECCV 2000. LNCS, vol. 1843, pp. 445–461. Springer, Heidelberg (2000)
57. Goshen-Meskin, D., Bar-Itzhack, I.Y.: Observability analysis of piece-wise constant systems - Part I: Theory. IEEE Transactions on Aerospace and Electronic Systems 28(4), 1056–1067 (1992)
58. Grosso, E., Metta, G., Oddera, A., Sandini, G.: Robust visual servoing in 3D reaching tasks. IEEE Transactions on Robotics and Automation 12(5), 732–742 (1996)
59. Guerrero, J.J., Murillo, A.C., Sagüés, C.: Localization and matching using the planar trifocal tensor with bearing-only data. IEEE Transactions on Robotics 24(2), 494–501 (2008)
60. Hartley, R.I., Zisserman, A.: Multiple View Geometry in Computer Vision, 2nd edn. Cambridge University Press (2004)
61. Hashimoto, K., Ebine, T., Kimura, H.: Visual servoing with hand-eye manipulator - Optimal control approach. IEEE Transactions on Robotics and Automation 12(5), 766–774 (1996)
62. Hashimoto, K., Kimura, H.: Visual servoing with nonlinear observer. In: IEEE International Conference on Robotics and Automation, pp. 484–489 (1995)
63. Hashimoto, K., Noritsugu, T.: Visual servoing of nonholonomic cart. In: IEEE International Conference on Robotics and Automation, pp. 1719–1724 (1997)
64. Hashimoto, K., Noritsugu, T.: Visual servoing with linearized observer. In: IEEE International Conference on Robotics and Automation, pp. 263–268 (1999)
65. Hermann, R., Krener, A.J.: Nonlinear controlability and observability. IEEE Transactions on Automatic Control 22(5), 728–740 (1977)
66. Hill, J., Park, W.T.: Real time control of a robot with a mobile camera. In: International Symposium and Exposition on Industrial Robots, pp. 233–246 (1979)
67. Hirschorn, R.M.: Output tracking through singularities. In: IEEE Conference on Decision and Control, pp. 3843–3848 (2002)
68. Huang, G.P., Mourikis, A.I., Roumeliotis, S.I.: Observability-based rules for designing consistent EKF SLAM estimators. The International Journal of Robotics Research 29(5), 502–528 (2010)
69. Hutchinson, S., Hager, G.D., Corke, P.I.: A tutorial on visual servo control. IEEE Transactions on Robotics and Automation 12(5), 651–670 (1996)
70. Isidori, A.: Nonlinear Control Systems. Springer, Great Britain (1995)
71. Jung, D., Heinzmann, J., Zelinsky, A.: Range and pose estimation for visual servoing of a mobile robot. In: IEEE International Conference on Robotics and Automation, pp. 1226–1231 (1998)
72. Khalil, H.K.: Nonlinear Systems, 3rd edn. Prentice Hall (2001)
73. Kim, J.K., Kim, D.W., Choi, S.J., Won, S.C.: Image-based visual servoing using sliding mode control. In: SICE-ICASE International Joint Conference, pp. 4996–5001 (2006)
74. Kragic, D., Christensen, H.I.: Robust visual servoing. The International Journal of Robotics Research 22(10-11), 923–939 (2003)

75. Lamiroy, B., Espiau, B., Andreff, N., Horaud, R.: Controlling robots with two cameras: How to do it properly. In: IEEE International Conference on Robotics and Automation, pp. 2100–2105 (2000)
76. Lee, H., Utkin, V.: Chattering suppression methods in sliding mode control systems. Annual Reviews in Control 31(2), 179–188 (2007)
77. Lee, K.W., Wijesoma, W.S., Ibanez-Guzman, J.: On the observability and observability analysis of SLAM. In: IEEE/RSJ International Conference on Intelligent Robots and Systems, pp. 3569–3574 (2006)
78. Li, F., Xie, H.L.: Sliding mode variable structure control for visual servoing system. International Journal of Automation and Computing 7(3), 317–323 (2010)
79. Li, M.H., Hong, B.R., Cai, Z.S., Piao, S.H., Huang, Q.C.: Novel indoor mobile robot navigation using monocular vision. Engineering Applications of Artificial Intelligence 21(3), 485–497 (2008)
80. Lippiello, K., Siciliano, B., Villani, L.: Visual motion estimation of 3D objects: An adaptive extended Kalman filter approach. In: IEEE/RSJ International Conference on Intelligent Robots and Systems, pp. 957–962 (2004)
81. Liu, Y.H., Wang, H., Wang, C., Lam, K.K.: Uncalibrated visual servoing of robots using a depth-independent interaction matrix. IEEE Transactions on Robotics 22(4), 804–817 (2006)
82. López-Nicolás, G., Gans, N.R., Bhattacharya, S., Guerrero, J.J., Sagüés, C., Hutchinson, S.: Homography-based control scheme for mobile robots with nonholonomic and field-of-view constraints. IEEE Transactions on Systems, Man, and Cybernetics - Part B: Cybernetics 40(4), 1115–1127 (2010)
83. López-Nicolás, G., Guerrero, J.J., Sagüés, C.: Multiple homographies with omnidirectional vision for robot homing. Robotics and Autonomous Systems 58(6), 773–783 (2010)
84. López-Nicolás, G., Guerrero, J.J., Sagüés, C.: Visual control of vehicles using two-view geometry. Mechatronics 20(2), 315–325 (2010)
85. López-Nicolás, G., Guerrero, J.J., Sagüés, C.: Visual control through the trifocal tensor for nonholonomic robots. Robotics and Autonomous Systems 58(2), 216–226 (2010)
86. López-Nicolás, G., Sagüés, C., Guerrero, J.J.: Homography-based visual control of nonholonomic vehicles. In: IEEE International Conference on Robotics and Automation, pp. 1703–1708 (2007)
87. López-Nicolás, G., Sagüés, C., Guerrero, J.J., Kragic, D., Jensfelt, P.: Switching visual control based on epipoles for mobile robots. Robotics and Autonomous Systems 56(7), 592–603 (2008)
88. Lowe, D.: Distinctive image features from scale-invariant keypoints. International Journal of Computer Vision 60(2), 91–110 (2004)
89. De Luca, A., Oriolo, G., Robuffo, P.: Image-based visual servoing schemes for nonholonomic mobile manipulators. Robotica 25(1), 131–145 (2007)
90. De Luca, A., Oriolo, G., Robuffo, P.: On-line estimation of feature depth for image-based visual servoing schemes. In: IEEE International Conference on Robotics and Automation, pp. 2823–2828 (2007)
91. De Luca, A., Oriolo, G., Samson, C.: Feedback control of a nonholonomic car-like robot. In: Laumond, J.P. (ed.) Robot Motion Planning and Control. LNCIS, vol. 229, pp. 171–253. Springer, New York (1998)
92. Lucas, B.D., Kanade, T.: An iterative image registration technique with an application to stereo vision. In: International Joint Conference on Artificial Intelligence, pp. 674–679 (1981)

93. Malis, E.: Visual servoing invariant to changes in camera-intrinsic parameters. IEEE Transactions on Robotics and Automation 20(1), 72–81 (2004)
94. Malis, E., Benhimane, S.: A unified approach to visual tracking and servoing. Robotics and Autonomous Systems 52(1), 39–52 (2005)
95. Malis, E., Chaumette, F.: Theoretical improvements in the stability analysis of a new class of model-free visual servoing methods. IEEE Transactions on Robotics and Automation 18(2), 176–186 (2002)
96. Malis, E., Chaumette, F., Boudet, S.: 2 1/2 D visual servoing. IEEE Transactions on Robotics and Automation 15(2), 234–246 (1999)
97. Marchand, E., Chaumette, F.: Feature tracking for visual servoing purposes. Robotics and Autonomous Systems 52(1), 53–70 (2005)
98. Marchand, E., Hager, G.D.: Dynamic sensor planning in visual servoing. In: IEEE International Conference on Robotics and Automation, pp. 1988–1993 (2000)
99. Mariottini, G.L., Oriolo, G., Prattichizzo, D.: Image-based visual servoing for nonholonomic mobile robots using epipolar geometry. IEEE Transactions on Robotics 23(1), 87–100 (2007)
100. Mariottini, G.L., Prattichizzo, D.: Image-based visual servoing with central catadioptric cameras. The International Journal of Robotics Research 27(1), 41–56 (2008)
101. Mariottini, G.L., Prattichizzo, D.: EGT: a toolbox for multiple view geometry and visual servoing. IEEE Robotics and Automation Magazine 12(4), 26–39 (2005)
102. Martinelli, A., Siegwart, R.: Observability analysis for mobile robots localization. In: IEEE/RSJ International Conference on Intelligent Robots and Systems, pp. 1471–1476 (2005)
103. Martinet, P., Gallice, J.: Position based visual servoing using a non-linear approach. In: IEEE/RSJ International Conference on Intelligent Robots and Systems, pp. 531–536 (1999)
104. Masutani, Y., Mikawa, M., Maru, N., Miyazaki, F.: Visual servoing for non-holonomic mobile robots. In: IEEE/RSJ International Conference on Intelligent Robots and Systems, pp. 1133–1140 (1994)
105. Mei, C., Rives, P.: Single view point omnidirectional camera calibration from planar grids. In: IEEE International Conference on Robotics and Automation, pp. 3945–3950 (2007)
106. Mezouar, Y., Abdelkader, H.H., Martinet, P., Chaumette, F.: Central catadioptric visual servoing from 3D straight lines. In: IEEE/RSJ International Conference on Intelligent Robots and Systems, pp. 343–349 (2004)
107. Mezouar, Y., Chaumette, F.: Path planning for robust image-based control. IEEE Transactions on Robotics and Automation 18(4), 534–549 (2002)
108. Mezouar, Y., Malis, E.: Robustness of central catadioptric image-based visual servoing to uncertainties on 3D parameters. In: IEEE/RSJ International Conference on Intelligent Robots and Systems, pp. 343–349 (2004)
109. Michel, O.: Webots: Professional mobile robot simulation. Journal of Advanced Robotics Systems 1(1), 39–42 (2004)
110. Murillo, A.C., Sagüés, C., Guerrero, J.J., Goedemé, T., Tuytelaars, T., Van Gool, L.: From omnidirectional images to hierarchical localization. Robotics and Autonomous Systems 55(5), 372–382 (2007)
111. Murrieri, P., Fontanelli, D., Bicchi, A.: A hybrid-control approach to the parking problem of a wheeled vehicle using limited view-angle visual feedback. The International Journal of Robotics Research 23(4-5), 437–448 (2004)
112. Nijmeijer, H.: Observability of autonomous discrete time nonlinear systems: A geometric approach. International Journal of Control 36(5), 867–874 (1982)

113. Nistér, D.: An efficient solution to the five-point relative pose problem. IEEE Transactions on Pattern Analysis and Machine Intelligence 26(6), 756–770 (2004)
114. Piazzi, J., Prattichizzo, D.: An auto-epipolar strategy for mobile robot visual servoing. In: IEEE/RSJ International Conference on Intelligent Robots and Systems, pp. 1802–1807 (2003)
115. Piepmeier, J.A., McMurray, G.V., Lipkin, H.: Uncalibrated dynamic visual servoing. IEEE Transactions on Robotics and Automation 20(1), 143–147 (2004)
116. Pomet, J.B.: Explicit design of time-varying stabilizing control laws for a class of controllable systems without drift. Systems and Control Letters 18(2), 147–158 (1992)
117. Reif, K., Unbehauen, R.: The extended Kalman filter as an exponential observer for nonlinear systems. IEEE Transactions on Signal Processing 47(8), 2324–2328 (1999)
118. Rives, P.: Visual servoing based on epipolar geometry. In: IEEE/RSJ International Conference on Intelligent Robots and Systems, pp. 602–607 (2000)
119. Royer, E., Lhuillier, M., Dhome, M., Lavest, J.M.: Monocular vision for mobile robot localization and autonomous navigation. International Journal of Computer Vision 74(3), 237–260 (2007)
120. Sagüés, C., Guerrero, J.J.: Visual correction for mobile robot homing. Robotics and Autonomous Systems 50(1), 41–49 (2005)
121. Sagüés, C., Murillo, A.C., Guerrero, J.J., Goedemé, T., Tuytelaars, T., Van Gool, L.: Localization with omnidirectional images using the radial trifocal tensor. In: IEEE International Conference on Robotics and Automation, pp. 551–556 (2006)
122. Samson, C.: Time-varying feedback stabilization of car-like wheeled mobile robots. The International Journal of Robotics Research 12(1), 55–56 (1993)
123. Samson, C., Le Borgne, M., Espiau, B.: Robot Control: The Task Function Approach. Oxford Engineering Science Series. Clarendon Press, Oxford (1991)
124. Sastry, S.: Nonlinear Systems: Analysis, Stability and Control. Springer, New York (1999)
125. Schramrn, F., Morel, G., Micaelli, A., Lottin, A.: Extended-2D visual servoing. In: IEEE International Conference on Robotics and Automation, pp. 267–273 (2004)
126. Shademan, A., Jagersand, M.: Three-view uncalibrated visual servoing. In: IEEE/RSJ International Conference on Intelligent Robots and Systems, pp. 6234–6239 (2010)
127. Shademan, A., Janabi-Sharifi, F.: Sensitivity analysis of EKF and iterated EKF pose estimation for position-based visual servoing. In: IEEE Conference on Control Applications, pp. 755–760 (2005)
128. Sharma, R., Hutchinson, S.: Motion perceptibility and its application to active vision-based servo control. IEEE Transactions on Robotics and Automation 13(4), 607–617 (1997)
129. Slotine, J.J.E., Li, W.: Applied Nonlinear control. Prentice-Hall, Englewood Cliffs (1991)
130. Svoboda, T., Padjla, T.: Epipolar geometry for central catadioptric cameras. International Journal of Computer Vision 49(1), 23–37 (2002)
131. Tahri, O., Chaumette, F.: Point-based and region-based image moments for visual servoing of planar objects. IEEE Transactions on Robotics 21(6), 1116–1127 (2006)
132. Tahri, O., Chaumette, F., Mezouar, Y.: New decoupled visual servoing scheme based on invariants from projection onto a sphere. In: IEEE International Conference on Robotics and Automation, pp. 3238–3243 (2008)
133. Thirthala, S., Pollefeys, M.: The radial trifocal tensor: A tool for calibrating the radial distortion of wide-angle cameras. In: IEEE Computer Society Conference on Computer Vision and Pattern Recognition, pp. 321–328 (2005)

134. Thirthala, S., Pollefeys, M.: Trifocal tensor for heterogeneous cameras. In: Proc. of 6th Workshop on Omnidirectional Vision, Camera Networks and Non-classical Cameras, OMNIVIS (2005)
135. Thuilot, B., Martinet, P., Cordesses, L., Gallice, J.: Position based visual servoing: keeping the object in the field of view. In: IEEE International Conference on Robotics and Automation, pp. 1624–1629 (2002)
136. Tsakiris, D., Rives, P., Samson, C.: Extending visual servoing techniques to nonholonomic mobile robots. In: Kriegman, D.J., Hager, G.D., Morse, A.S. (eds.) The Confluence of Vision and Control. LNCIS, vol. 237, pp. 106–117. Springer, Heidelberg (1998)
137. Usher, K., Ridley, P., Corke, P.: Visual servoing of a car-like vehicle - an application of omnidirectional vision. In: IEEE International Conference on Robotics and Automation, pp. 4288–4293 (2003)
138. Guldner, J., Utkin, V., Shi, J.: Sliding Mode Control in Electromechanical Systems. CRC Press, Boca Raton (1999)
139. Vidal-Calleja, T., Bryson, M., Sukkarieh, S., Sanfeliu, A., Andrade-Cetto, J.: On the observability of bearing-only SLAM. In: IEEE International Conference on Robotics and Automation, pp. 4114–4119 (2007)
140. Wang, H., Liu, Y.H., Zhou, D.: Adaptive visual servoing using point and line features with an uncalibrated eye-in-hand camera. IEEE Transactions on Robotics 24(4), 843–857 (2008)
141. Weiss, L., Sanderson, A., Neuman, C.: Dynamic sensor-based control of robots with visual feedback. IEEE Journal of Robotics and Automation 3(5), 404–417 (1987)
142. Wilson, W.J., Hulls, C.C.W., Bell, G.S.: Relative end-effector control using cartesian position-based visual servoing. IEEE Transactions on Robotics and Automation 12(5), 684–696 (1996)
143. Yu, Y.K., Wong, K.H., Chang, M.M.Y., Or, S.H.: Recursive camera-motion estimation with the trifocal tensor. IEEE Transactions on Systems, Man, and Cybernetics - Part B: Cybernetics 36(5), 1081–1090 (2006)
144. Zanne, P., Morel, G., Plestan, F.: Robust vision based 3D trajectory tracking using sliding mode. In: IEEE International Conference on Robotics and Automation, pp. 2088–2093 (2000)
145. Zhang, H., Ostrowski, J.P.: Visual motion planning for mobile robots. IEEE Transactions on Robotics and Automation 18(2), 199–208 (2002)

Kolibri 360 GmbH
Ferdinand-Jühlke-Straße 7
99095 Erfurt, Deutschland
produktsicherheit@kolibri360.de